Moon
Mars and Venus

A CONCISE GUIDE IN COLOUR

Moon
Mars and Venus

Antonín Rükl

HAMLYN

London · New York · Sydney · Toronto

Illustrations on endpapers: *front* — schematic map of the Moon
back left — general map of both hemispheres of Mars with names of the most prominent albedo formations
back right — albedo formations of Mars visible through telescopes with an aperture of 15 cm and more

Consultant Editor Dr. R. E. W. Maddison, F. S. A.
Translated by Daniela Coxon

Designed and produced by Artia for
The Hamlyn Publishing Group Limited
London ● New York ● Sydney ● Toronto
Astronaut House, Feltham, Middlesex, England

Reprinted 1978

Copyright © 1976 Artia, Prague

All rights reserved.
No part of this publication may be reproduced or transmitted in any form or by any means, electronic or mechanical, including photocopy, recording, or any information storage and retrieval system, without permission in writing from the copyright owner.

ISBN 0 600 36219 1

Printed in Czechoslovakia
3/02/24/51-02

Contents

Introduction   7

THE MOON   8

  The Moon in the Sky   8
  Near and Distant Views of the Surface of the Moon   12
  Lunar Charts   19
  Equipment and Man on the Moon   24
  Lunar Probes   29

Discovery of the Planets   33

VENUS   35

  The Evening and the Morning Star   35
  Venus Through the Telescope   36
  What We Know about Venus   38
  Probes to Venus   42

MARS   45

  Orbit Around the Sun   45
  Mars Through the Telescope   46
  Mapping of the Planet   50
  A Close View of Mars   52
  Satellites of Mars   56
  Probes to Mars   57

Explanatory Notes for Maps of the Moon and Mars   60

GENERAL MAP OF THE MOON   64—75

DETAILED MAP OF THE NEAR SIDE OF THE MOON   76 — 227

MAP OF MARS   228 — 239

A Comparative Table of Data on the Moon and Planets   241

Moon Craters Named Since 1973   241

Index to the Maps of the Moon   245

Index to the Maps of Mars   252

Acknowledgments   255

Introduction

The development of astronautics at the end of the 1950s led to the introduction of completely new techniques of research in the field of astronomy. Interplanetary flights ceased to be merely products of the imagination. By the 1960s revolutionary changes were taking place in the long history of astronomy as bodies of the solar system increasingly became the object of direct research. The first of these was the Moon followed closely by our two neighbouring planets, Venus and Mars.

Pioneer research into the immediate cosmic neighbourhood of the Earth precipitated an unprecedented interest in the Moon and adjacent planets, as the detailed study of these bodies provides the key to a real understanding of the Earth's past history which attracts the interest of a whole range of scientists. There is also the vast body of amateur astronomers, who devote much of their leisure time to the study of the universe, and who quite rightly feel that they are directly involved in this great work of investigation. It is largely for their benefit that this book has been produced, although it may also serve as a reference work for other scientists, who are not specifically astronomers and have never studied the Moon and the planets closely, but who have suddenly found that astronautics has brought these bodies within the scope of their field of study.

However, the aim of this pocket atlas is basically a modest one, namely, to present to the reading public such details of phenomena pertaining to the Moon, Venus and Mars as can be observed from the Earth by the naked eye or through a telescope. The most useful section of the atlas contains detailed maps of the Moon, in which special attention has been paid to the nomenclature of the Moon's formations. The general and detailed maps of the Moon cover eighty-two colour plates and the remaining plates consist of a six-part map of Mars. The smallest part of this atlas has of necessity been allocated to the description of Venus because, apart from its well-known phases, only a negligible amount of detail of this planet can be observed through a telescope.

THE MOON

Since time immemorial the Moon has been held in esteem and respect. This is hardly surprising as the Moon is the most evident heavenly body apart from the Sun; it shines at night, and has always helped in the measurement of time. In addition, ancient astrologers attributed to the Moon certain characteristics believed to influence the nature of the Earth and of Man himself. Today's attitude to the Moon is rather more critical and is based simply on its physical activity, which can be objectively observed and measured in terms of its impact on the Earth. This involves such well-known phenomena as the ebb and flow of the tides, caused by the gravitational pull and motion of the Moon.

Interest in the Moon has greatly increased in recent years, being stimulated by the desire for a deeper understanding of the past and the probable future of our planet, the Earth. Such an understanding is directly related to better knowledge of adjacent celestial bodies and their own history. Herein lies the main justification and aim of all research centred on the Moon and the planets.

The Moon in the Sky

Viewed with the naked eye, the Moon has always shown the same image to observers on Earth. However, the Moon is now regarded in a completely different way from that, let us say, of Neolithic man. Although early man was an attentive observer of nature and superficially saw the same luminary as modern man, nevertheless for him the Moon remained only a strange and awesome silvery disc. The contemporary notion of the Moon is of a close celestial body covered with craters, which has no air, water or life. However, present knowledge of the Moon is far from being so generalized or limited in detail.

The Moon's orbit around the Earth is an elliptical one and therefore its distance from our planet is constantly changing. The point of closest approach to the Earth is called the *perigee*, and the point farthest from the Earth is called the *apogee*. When the Moon is at perigee, its distance from the Earth is 356 500 kilometres and it appears in the sky as a disc with a diameter of 33′30″. At apogee the Moon is 406 700 kilometres from the Earth and its apparent diameter is 29′22″. Explained in simpler terms the Moon can be seen in the sky as a disc with an apparent diameter of about half a degree and therefore equal in size to the solar disc. In consequence of the Moon's period of rotation on its axis being equal to its period of revolution around the Earth, the Moon always presents the same face to an observer on the Earth.

The appearance of the Moon in the sky depends on the respective positions of the Sun, the Moon and the Earth in space. Such changes in the position of the Moon are called its phases. Starting with the *New Moon*, namely when the Moon is between the Sun and Earth, the phases progress through the *first quarter* to the *Full Moon*, when the Earth is between the Moon and the Sun; then to the *third quarter* and back again to the New Moon. The time it takes the Moon to pass through all its phases is called a *synodic month* and lasts 29 days, 12 hours and 44 minutes. This constitutes one complete revolution of the Moon around the Earth in terms of its relationship to the Sun.

The New Moon can be seen only very rarely, when it is close to the line linking the Sun and the observer — that is, when an eclipse of the Sun occurs. The slender Crescent Moon appears in the evening sky only one day after the New Moon. It is usually visible from the Earth about 25 hours before or after the New Moon and only very occasionally can it be observed earlier than 20 hours before or after. During a period of 2 to 4 or even more days before and after the New Moon, this brightly shining Crescent Moon is also faintly illuminated by sunlight reflected from the Earth's surface, so that the Moon for some time has the appearance of a full circle.

The motion of the Moon relative to the background of the sky is very fast. In the space of 1 day it describes a curve of about 12° moving from the west to the east. The time required for the Moon to make one complete revolution with respect to a fixed star is called a *sidereal month* and lasts 27 days, 7 hours and 43 minutes. The Moon rises every day a little later than on the previous day, and its visibility above the horizon also changes. For example, the growing Crescent

Moon is visible in the evening, the Full Moon shines all night, whilst the waning Moon appears only after midnight. Similarly, the Moon's height above the horizon fluctuates sharply, just as the height of the Sun above the horizon changes during the course of the year.

The hemisphere of the Moon facing the Earth is covered with dark patches, which are instantly apparent to the naked eye. In many countries these patches are associated with the picture of a hare, a tortoise, a lizard or a woman carrying a basket on her back, or two faces kissing. Finally the smiling human face of the 'man in the Moon' still appears on some calendars as the Moon's mark of identity.

When astronomers in the seventeenth century began to observe the Moon through telescopes, they concluded that, as in the case of the Earth, it had seas and dry land. Since then the dark areas on the Moon have been called seas or *maria* (sing. mare) and the light areas called continents or *terrae* (sing. terra), although we know now that there is no water on the Moon. The first maps of the Moon dating from the mid-seventeenth century also featured lakes *(lacus)*, marsh *(palus)* and bays *(sinus)*. These names were introduced by Riccioli in 1651.

Similarly, it was once thought that the Moon influenced the weather on Earth according to a simple theory which maintained that the weather would be fine when the Moon was waxing and that it would be cloudy, rainy and stormy when the Moon was waning. Riccioli apparently had this idea in mind when he was inventing names for the maria. For example, those dark areas which are visible during the waxing Crescent leading to the first quarter were given names associated with fine weather, such as Sea of Tranquility and Sea of Serenity. Similarly the dark areas visible during the waning Moon from the third quarter onwards were given names associated with bad weather, such as Sea of Showers, Sea of Clouds, Ocean of Storms and Sea of Moisture.

The names of the maria provide the basis of lunar topography and because they are used as the basic points of orientation on the Moon, they warrant rather more attention. The accompanying map (Fig. 1) is a diagram of the Full Moon as it can be seen by the naked eye. In addition, three

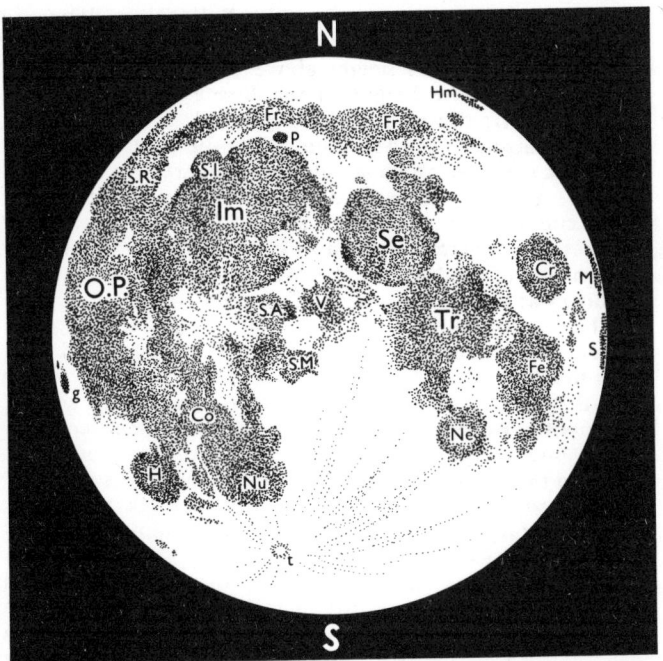

Fig. 1. *The map of lunar seas visible to the naked eye applied to a sight hole*

Co — Mare Cognitum (Known Sea) Cr — Mare Crisium (Sea of Crises) Fe — Mare Fecunditatis (Sea of Fertility) Fr — Mare Frigoris (Sea of Cold) H — Mare Humorum (Sea of Moisture) Hm — Mare Humboldtianum (Humboldt's Sea) Im — Mare Imbrium (Sea of Rains) M — Mare Marginis (Border Sea) Ne — Mare Nectaris (Sea of Nectar) Nu — Mare Nubium (Sea of Clouds) O.P. — Oceanus Procellarum (Ocean of Storms) Se — Mare Serenitatis (Sea of Serenity) S.A. — Sinus Aestuum (Bay of Billows) S.I. — Sinus Iridum (Bay of Rainbows) S.M. — Sinus Medii (Central Bay) S.R. — Sinus Roris (Bay of Dew) Tr — Mare Tranquillitatis (Sea of Tranquillity) V — Mare Vaporum (Sea of Vapours)
craters: g — Grimaldi p — Plato t — Tycho

prominent craters which are visible with the help of binoculars are also included.

The Moon always presents the same face to the Earth, because of the nature of its motion, as described above. The Moon is, therefore, arbitrarily divided into two hemispheres: the near side, visible from Earth; and the far side, invisible from the Earth. The borderline between the two hemispheres is, however, a changeable one. To the terrestrial observer it seems as if the Moon is slowly rocking about a central position. This oscillation is called *libration* and arises from irregularity superimposed on the Moon's rotation. There are three causes, two of which are uneven orbiting speed of the Moon (i.e. libration in longitude, which can cause a displacement of $\pm 7°5'$); and the inclination of the rotating axis of the Moon (i.e. libration in latitude within limits of $\pm 6°50'$). Thus, on account of libration it is possible to observe almost 59 per cent of the Moon's surface from the Earth. This means that 41 per cent of the Moon's surface can never be seen from the Earth, and the 18 per cent of the lunar surface, which is called the libration zone, can be observed only at certain times.

Near and Distant Views of the Surface of the Moon

The Moon is the only celestial body which offers such a wealth of observable details. Observations can be conducted with the aid of only a small telescope, if it has a good optical system. For example, a telescope with an objective of $5-10$ centimetres in diameter can be used to see the majority of formations marked in this atlas on maps $1-76$. However, a telescope with a superior objective of $20-25$ centimetres in diameter can reveal details smaller than 1 kilometre in diameter, the same lunar details as are photographed by the largest telescopes in the world. Additionally, the quality and amount of detail depend on the stability of the telescope's mounting, on atmospheric conditions and on the experience of the observer.

The Moon can be observed in greatest detail along the *terminator*, which is the boundary between day and night, where the Sun rises (i.e., the morning terminator) or where the Sun sets (the evening terminator). The angle of incidence of the Sun's rays is then very small, and the shadows cast on the Moon's surface are long. The first viewing through even a small telescope will reveal the striking differences between the dark and light areas on the Moon's surface. The dark areas of the maria appear largely the same as they do to the naked eye. However, the light areas under the telescope reveal a number of details, of which craters are the predominant feature and one of the most characteristic formations. The observable face of the Moon is covered with about 300 000 craters with a diameter over 1 kilometre. The largest of these are over 300 kilometres in diameter and the smallest ones visible from the Earth measure about 500 metres across. The camera of an automatic lunar probe will detect at short distances a great many craterlets, which vary considerably in the size of their diameter. However, the smallest Moon craters were discovered during examinations of lunar rock samples under a microscope on the Earth. It follows then that the term 'crater' has a wide meaning and is used to describe circular formations of various shape, size and origin. The following are the basic types of craters.

The largest craters whose diameters range from 60—300 kilometers are *walled plains*. As the name suggests these are relatively flat areas surrounded by a circular wall. The floor of the plain is not always flat and smooth, but is often covered with smaller craters, hill formations and rilles. The wall is usually fairly high, noticeably segmented and often broken up by furrows, clefts, small craters and landslides. Among the most typical walled plains are Clavius (p. 218), Ptolemaeus (p. 162), Schickard (p. 198) and Posidonius (p. 102).

Figure 2 shows several typical lunar formations, and includes a walled plain (1). The floor of the walled plain usually lies slightly below the level of the surrounding surface and has the same curvature as the rest of the Moon's surface. For example, the diameter of Ptolemaeus is 153 kilometres and its depth measured from the top of the surrounding wall to the base of the plain is as much as 2.4 kilometres so that the ratio of the height to the diameter is 1 : 64. Therefore, an astronaut who might be standing on the plain several dozen kilometres from the foot of the wall, would not be aware that he is in a crater and in many

Fig. 2. *Typical formations of the lunar surface*
T — continent (terra)   mm — dark mare material   v — wall of a crater   M — sea (mare)   c — central peak   1 — walled plain   2 — circular mountain range   3 — crater with a flooded floor   4 — crater with a sharp rim   5 — valley (vallis)   6 — small crater   7 — flooded crater   8 — mountain range (montes)   9 — mountain (mons)   10 — dome   11 — wrinkle ridge   12 — fault (rupes)   13 — sinuous rille   14 — rille, cleft
1', 2', ... 14' ideal profiles of formations shown
Data in kilometres give typical crater diameters.
Typical heights are given in metres.

instances would not be able to see the top of the surrounding wall because of the considerable curvature in the surface of the Moon.

The distance of the horizon d is given by the formula
$$d = \sqrt{D \times h},$$
where D is the diameter of the Moon and h is the height of the observer's eye above the Moon's surface.

For example, if h = 1.6 m = 0.0016 km (i.e., the eye level of an upright man) then
$$d = \sqrt{3500 \times 0.0016}$$
$$= 2.4 \text{ km}.$$
This means that in a flat region of the Moon it is possible to see to a distance of 2.4 km. Similarly, it can be estimated at what point a hill or a wall would disappear below the horizon from an observer's view.

For example, if the wall's height is 1 000 m, then
$$d = \sqrt{3500 \times 1} + 2.4 \text{ km}$$
$$= 59.2 + 2.4 \text{ km}$$
$$= 62 \text{ km}.$$

The exact formula is $d = \sqrt{D \times h + h^2}$.

Perhaps the most beautiful craters with a considerable variety of features are the *ring mountains*, such as Copernicus (p. 136), Theophilus (p. 166), and Tycho (p. 202). The diameter of such ring mountains ranges from 20—100 kilometres. Their regular, enclosing, circular wall has a relatively sharp crest, the inner slopes of which are terraced with

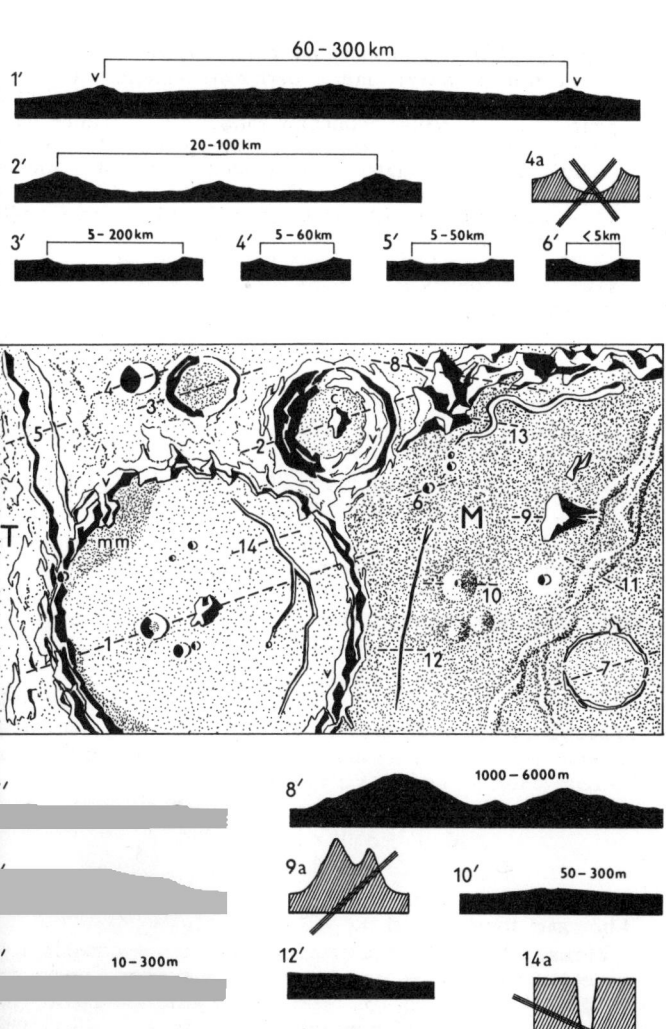

a gradient of 20° to 30°, whilst the outer slopes have a more gradual gradient of 5° to 15°. The floor of the ring crater often has one or several central peaks rising from it, or it may be interspersed with craters, rilles or other formations of that kind, and is lower than the surrounding terrain on the outside. According to Schröter's law, the volume of a wall is equivalent to the spacial content of the crater depression. A typical profile of a ring mountain (2') in Fig. 2 illustrates a very shallow formation.

Circular depressions with a wall of 5—60 kilometres in diameter are called *craters*. A typical crater as shown in Fig. 2 (4') has a single circular wall with a relatively sharp crest, having an unterraced inner wall and no central peaks. The ratio of the depth to the diameter is 1 : 5. The profile (4') shows one such relatively deep crater (e.g. the crater Hortensius (p. 134), which has a diameter of 14.6 kilometres and a depth of 2.86 kilometres).

The reader who comes across such information for the first time will probably find it difficult to believe that craters are not deep hollows with steep slopes (illustrated incorrectly in profile in Fig. 2, 4a). Under conditions of oblique illumination the long, black shadows create an illusion of great height of the formation which casts the shadow. This false impression is particularly strong in the case of lunar mountains.

The smallest craters visible from the Earth are called *craterlets* and *crater pits*, and their diameter ranges from 5 kilometres downwards. The text accompanying the detailed maps of the Moon in this book contains much detailed, statistical data on the size of both small and large craters. Such data can even prove helpful to those observers who merely want to test the resolving power of their telescope.

Craters come in various shapes and sizes. Their walls, for example, need not be always circular, for many are polygonal and on rare occasions even concentric double walls can be found (e.g. Hesiodus A, Marth). Additionally, many crater floors have been flooded by lava, but only exceptionally is an entire crater filled up to the top of its ramparts with volcanic material (e.g. the crater Wargentin, p. 214). In lunar seas, under very oblique illumination, it is just possible to see 'ghost' craters — i.e., craters completely filled with lava.

The relative steepness of the walls, the extent of their disintegration, the infill of the crater floor and the merging of adjacent craters, as well as many other features, are factors determining the relative age of craters. However, the problem of the origin of lunar craters is a much more difficult one. An analysis of the problems arising from a study of the origin of such lunar formations would occupy too much space and would in any case be largely hypothetical, so only two basic propositions will be considered:

a. that craters have developed from endogenous forces, such as volcanism in its widest sense, and tectonic and other processes.

b. that craters have developed from exogenous causes, such as explosions following the impact of meteorites on the Moon's surface.

It is evident that the lunar surface has been determined by both internal and external types of activity, and thus the views of selenologists have developed in two ways. Volcanists maintain that at least 95 per cent of the formations developed as a result of endogenous activity, and perhaps the remaining 5 per cent was caused by the impact of meteorites or of rocks thrown out by volcanic activity. On the other hand, scientists who support the impact theory assume that such surface formations resulted largely from the impact of meteorites or the secondary impact of lunar material following primary explosions. Such hypotheses regard volcanic activity as simply a process triggered off by such impacts. The theory of the impact origin of the majority of craters has been the predominant view held by members of those international conferences that met to consider the results of the Apollo expeditions.

However, for every formation it is necessary to consider separately the possibility of its volcanic or impact origin. This is no simple problem and a decision cannot be arrived at merely on the basis of analogy between a lunar crater and a specific crater on Earth which is thought to have a certain origin. The Moon is a completely different celestial body from the Earth, with a distinctive internal structure, gravity and magnetism as well as a different history.

The *mountain ranges* on the Moon are also different and although their names are terrestrial (e.g. the Alps, the

Caucasus, the Carpathians), their character is truly lunar. The characteristic network of valleys is absent on the Moon's surface, as these are formed between the Earth's mountains by water erosion. Fig. 2 (8′) shows a profile of a lunar range of mountains and Fig. 2 (9′) a profile of an isolated mountain, such as Piton or Pico. These are not needle-like peaks, with steep slopes and sharp edges, nor are they spiky peaks (c.f. incorrect profile (9a)). They have gradual slopes with a gradient of between 15° to 35°, which form rounded mountain ranges, and look as though they were modelled out of sand. The most typical formations of endogenous origin are the lunar *domes*. Good examples of such domes are near the crater Hortensius. Their diameter is usually between 10−20 kilometres, their height is several hundred metres, and their slopes are of the order of 1° to 3°, which means they are visible only close to the terminator. In terms of the gradient of their slopes, height and visibility, such domes resemble the so-called marial mountain ridges (wrinkle ridges), which characterize the lunar seas. In appearance they most resemble prominent veins.

Other interesting objects for telescopic observation include *faults* (e.g. Rupes Recta, p. 182), *rilles* or *clefts* (Rimae Triesnecker, Rima Hyginus, pp. 140, 142) and *sinuous rilles* (Vallis Schröteri, p. 110).

In view of the historical development of the Moon some attention must be paid to those sizeable circular depressions called *basins*, with their conspicuous concentric walls and radial structures. These formations are related to the large walled plains and circular maria. Some of these, for example Mare Imbrium and Mare Orientale, have a very pronounced system of streaks, which radiate in all directions from some of the depressions in the form of long furrows, mountain ranges and valleys for hundreds of kilometres across the lunar surface.

The mare areas are perhaps one of the greatest lunar puzzles. They are formed from dark 'mare' material and it is generally considered that they developed as a result of a gradual flowing of lava over vast plains. However, the surfaces of these large marial areas are not horizontal and do not follow the surface of a perfect sphere, but are often inclined as much as several degrees and have an irregular undulating surface.

The best visibility of various lunar features is in proximity to the terminator. When the shadows disappear, some craters merge with their greyish background, whilst others stand out as clear points, plains or rings. Some peaks and clefts also shine brightly at this juncture. At Full Moon, bright rays extend outwards from numerous craters, such as Tycho, Copernicus and Kepler. These rays probably consist of ejecta.

To achieve the best understanding of the lunar surface some knowledge of the physical conditions of the Moon is necessary. Since its gravitation is one-sixth that of the Earth, the Moon is incapable of maintaining an atmosphere of gas at its surface. There is also no water on the Moon, neither in a free form nor incorporated in its surface minerals. The lunar day lasts for 29.5 terrestrial days, which equals 708 hours; the temperature during this period fluctuates between $+ 110°C$ at noon to $- 170°C$ at the end of night. This large fluctuation in the temperature is, however, limited to the surface layer of loose material, which envelops the whole lunar surface from the marial areas to the continents. This layer of material is many metres thick and consists of fragmented material, such as large stones, dust and mineral particles, which have resulted from numerous explosions that took place several thousand million years ago, when the compact rock was disintegrated.

Lunar Charts

Until 1959 the mapping of the Moon was limited to its near side, and consequently the vast majority of the resulting maps presented the Moon in the same way as the terrestrial observer sees it in the sky. However, the Soviet probe, Luna 3, heralded a new epoch of global cartography of the Moon's surface. The whole Moon was mapped in detail during the Lunar Orbiter programme in 1966—67. As a result of this exploration the later Apollo expeditions were equipped with highly detailed maps.

To begin mapping the Moon, knowledge of its size and shape is obviously necessary. The study of these problems is known as *selenodesy*. Selenodetic measurements have indicated that the Moon has a spherical shape with a mean radius of 1 738 kilometres. Its surface, however, is irregular, and local deviations from the mean semidiameter are between 2 and 4 kilometres. The more practical problems of the Moon's topography are studied by *selenography*. This deals with the selection of scales and the associated planning of charts, the cartographical determination of the exact location of lunar formations, the determination of the relative heights of mountains and depths of craters and finally the nomenclature to be used.

The development of the nomenclature used in lunar charts needs further consideration, because this detailed nomenclature forms the main content of maps 1 — 76 in this book. The first names were given to the lunar formations by the makers of the earliest charts in the middle of the seventeenth century. The foundations of the current system of nomenclature were laid by the Jesuit, Giovanni Battista Riccioli, a Professor of Philosophy, Theology and Astronomy in Bologna, who in 1651 published a work called *Almagestum Novum*. This book contains a map of the Moon with the names given to the lunar formations by the author himself. He based his nomenclature on maps of previous observers. One of these was a Polish astronomer, Hevelius of Danzig, from whom he took the idea of naming the lunar mountain ranges after the mountain ranges on the Earth, and also of calling the dark areas on the Moon maria. Using the example of Langren's map of 1645, Riccioli named craters after contemporary, medieval and ancient astronomers. Lunar nomenclature was further developed by the German scientists Schröter *(Selenotopographische Fragmente*, 1791 and 1802*)*, Beer and Mädler *(Der Mond*, 1837*)*. They not only supplemented Riccioli's system by adding many new names, but also introduced a system of naming small craters by the capital letters of the Latin alphabet and designating the mountains by lower case Greek letters.

However, the first internationally accepted nomenclature was agreed as late as 1935. This uniform nomenclature, containing 672 names, was compiled by an English scientist, M. A. Blagg, and a German scientist, K. Müller, who were officially entrusted with this task by the International Astronomical Union (IAU), and who published the product of their joint endeavours in 1935 under the title *Named*

Lunar Formations: Catalogue and Map. Since that time the IAU has been the sole authority that has presided over all changes in and additions to lunar terminology.

Further development in lunar nomenclature took place at the beginning of the 1960s and was associated with the growth of the lunar probe programmes and the subsequent mapping projects. In 1961 the IAU accepted the first names given to the formations on the far side of the Moon, which was photographed by Luna 3. An additional innovation introduced at the XII General Assembly of the IAU in 1964, apart from the acceptance of sixty-six new names, was the general latinization of all terms used in naming topographical formations (e.g. *mons* = mount, *montes* = mountains, *rupes* = fault, *rima* = rille or cleft, *vallis* = valley).

At the XIV General Assembly of the IAU in Brighton in 1970 more new terms, amounting to 513 in all, were accepted for the far side of the Moon. Then an unusual reform was introduced at the XV General Assembly in Sydney, when it was realized that the cartographical system in use did not meet the requirements of detailed mapping (for scales of 1 : 250 000 and upwards) and the designation of a large number of small, secondary craters by groups of letters was unclear and unsatisfactory. A proposal for a new system was therefore accepted. This was based on the idea of subdividing the Moon by a network of coordinates into 144 regions identical with the layout of the map LAC 1 : 1 000 000 (NASA Lunar Aeronautical Chart). Each region is further divided into sixteen sections designated by combinations of letters A, B, C, D, and numbers 1, 2, 3, 4. Apart from this, every section carries the name of some local crater. One problem caused by this system is the lack of a sufficient number of named craters in these sections, since even the near side of the Moon has approximately another 500 craters to be named. The first step in solving this problem was taken in 1973 when the IAU accepted 53 new names, out of which 43 were ascribed to those craters located in the equatorial zone on the near side of the Moon, where the detailed mapping process has already started (see list on pp. 241 — 243). In the near future, therefore, the old system of indicating associated groups of craters by letters will be gradually phased out and such craters will receive individual names. All in all about

10 000 new names are needed for the whole face of the Moon. However, it is not yet absolutely clear whether this idea will prove to be practical. The traditional system of marking peaks and mountains by Greek letters will also be replaced by suitable names. Additionally, new Latin terms are being introduced, such as *dorsum* (pl. *dorsa*) for the ridges, *fossa* (pl. *fossae* = ditches) for the flat ditch-like furrows, *anguis* (pl. *angues* = snakes) for the sinuous rilles and *catena* (pl. *catenae* = chains) for the crater chains.

The practice of choosing the names of prominent deceased scientists as well as writers, painters and musicians to provide lunar names is going to be maintained. However, political, military and religious personalities as well as modern philosophers have been excluded from taking place in this developing lunar pantheon of personalities.

All the details of these new reforms cannot possibly be mentioned, as it is a far-reaching project which will not be accomplished for many years. Before these ventures are realized, new, very detailed maps of the whole surface of the Moon will have to be compiled and new catalogues of the names, specifications and coordinates of lunar formations will have to be published, incorporating comparative tables of the old and new specifications. Obviously, therefore, the old system will be in use for many years, if not decades to come. In any case general maps of the Moon designed for the amateur will not be affected very much by this change because the reforms of 1973 mainly concern detailed maps.

The names of lunar formations are a practical guide for quick orientation. The exact and explicit data for a certain point on the Moon require selenographic coordinates, which are similar to geographical coordinates. Figure 3 shows a network of such lunar coordinates as seen from the Earth. The basic circle of the system of selenographic coordinates is the lunar equator e, which lies on a plane perpendicular to the axis of rotation a. The axis of rotation intersects the surface of the Moon at the North Pole N and the South Pole S. Orientation to the east and west is based on a decision of the IAU in 1961 and is defined in the sense of the observer on the Moon, for whom the Sun rises naturally in the east and sets in the west. When looking at the Moon with the naked eye or through a terrestrial telescope, the north (N) is at the top, the south (S) at the bottom, the east (E) on the right in the direction of Mare Crisium and the west (W) on the left in the direction of Oceanus Procellarum. This orientation is used in all modern maps of the Moon's surface.

Selenographic longitude (λ = lambda) is the angle between the central meridian (m) and the meridian containing the given point. It is measured positively to the east and negatively to the west, always from 0° to 180°. The abbre-

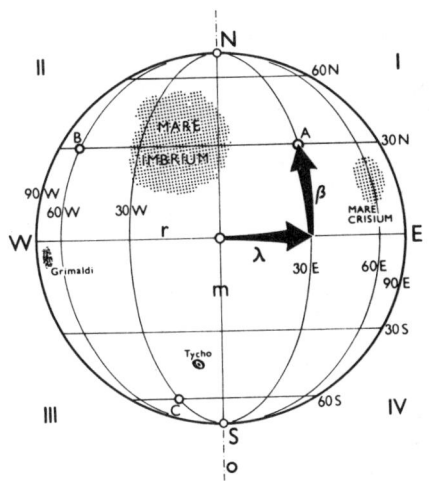

Fig. 3. *Selenographic coordinates*
N — North E — East e — equator a — axis of rotation of the Moon S — South W — West m — prime meridian
λ — selenographic longitude β — selenographic latitude
Coordinates of points A, B and C are: A (30 E, 30 N)
B (60 W, 30 N)
C (30 W, 60 S)

I, II, III and IV ... quadrants

viations E for east and W for west are often used instead of signs + and −. For example, 30W signifies 30° of western selenographic longitude ($\lambda = -30°$).

Selenographic latitude (ß = beta) is the angular distance of the given point from the equator, reckoned on the given meridian positively to the north and negatively to the south, always from 0° to 90°. The abbreviations N for north and S for south are often used instead of the signs + and −. For example, 60S means 60° of southern selenographic latitude ($\beta = -60°$).

An important datum for the observer of the Moon is *co-longitude*, which is the selenographic longitude of the morning terminator (see p. 12), measured from the central meridian westwards from 0° to 360°. The values of co-longitude for every day are listed in astronomical year-books.

Equipment and Man on the Moon

This small guide documents the pioneering era in the history of lunar research. The numerous maps record the places of landing of the lunar probes and the Apollo expeditions, the exciting voyages of discovery of which will forever remain in human memory.

The beginnings of space exploration were subject to two very different attitudes and two long-term programmes, namely the Soviet and the American, which developed independently of each other.

The Soviet programme consisted of three series of Luna probes. The simple probes of the first series bypassed the Moon (Luna 1, 1959), then for the first time crash-landed on the Moon (Luna 2, 1959) and finally photographed a part of the far side of the Moon (Luna 3, 1959). The probes of the second series (Luna 5 to Luna 14, 1965—68), after several abortive attempts, achieved the first soft landing on the Moon (Luna 9, 1966) and established the first artificial lunar satellite (Luna 10, 1966). Of specific importance in the Soviet programme were the Zond probes, which orbited the Moon, took photographs and measurements from close distances and in some cases returned to the Earth (Zond 3, 1965; Zonds 6—8, 1968—70). Finally the probes belonging to the third series were inaugurated by Luna 15 in 1969.

Its basic form consisted of a multipurpose transport platform, which made a soft landing on the Moon by means of a four-legged undercarriage. This landing section transported to the Moon various types of useful equipment, such as the Moon-Earth rocket equipped for the automatic collection of rock samples (Luna 16, 1970 and Luna 20, 1972), or with a remote-controlled mobile laboratory (Luna 17 with vehicle Lunokhod 1, 1970 and Luna 21 with Lunokhod 2, 1973).

The American astronautical, lunar research programme began with a series of experiments involving the Ranger probes, which were directed to crash-land in pre-selected areas of the Moon's surface. Six cameras photographed the Moon shortly before this final descent and these photographs were immediately transmitted to the Earth (Ranger 7—9, 1964—65). Between 1966 and 1968 two simultaneous programmes were carried out in preparation for the landing of astronauts on the Moon. They involved direct research into the nature of the lunar surface by the soft-landing Surveyor probes and the mapping of the Moon by the Lunar Orbiter artificial satellites. The Surveyor probes were equipped with a television camera with a zoom-lens, remotely controlled from the Earth, a mechanical surface-sampler (Surveyor 3 and 7) and facilities for chemical analysis using alpha-particle scattering (Surveyor 5, 6 and 7). Five successful Surveyor probes transmitted a total of 87 674 photographs to the Earth.

Similarly, some very important work was completed by the five Lunar Orbiter satellites. Their original task was to photograph the areas chosen for the Apollo expeditions. However, this task was soon completed by the first three Orbiters and therefore their programme was extended to a global mapping of the Moon's surface. The American programme finally culminated in a series of manned flights to the Moon, known as the Apollo programme. The first were orbital flights around the Moon (Apollo 8, 21—27 December 1968 and Apollo 10, 20—26 May 1969). These were followed by the first manned landing on the Moon (Apollo 11, 20 July 1969) and another five similar expeditions (Apollo 12, 14, 15, 16 and 17 between 1969 and 1972). The main task of these expeditions was geological survey work, the collection of rock samples, photographical documentation and the installation of a geophysical station on the Moon's surface. Extremely valuable information was also obtained throughout by the apparatus located in the command section of the Apollo spaceship orbiting the Moon.

These explorations have produced many surprises, sensations and real discoveries, which can be roughly summarized as follows:

1. The whole of the surface of the Moon is covered by a layer of dust several centimetres thick. However, the surface is firm and the astronaut's boots or the wheels of the electrically powered vehicle only sink into it a few centimetres. Prior to the landing of Luna 9 on the Moon, it was thought that this layer of very fine dust was many metres thick, and that all solid bodies would sink into it completely.

2. The atmospheric pressure on the diurnal side of the Moon is approximately 10^{-13} of the pressure near the Earth's surface. It was discovered that during the day about a million atoms of hydrogen and helium and 60 000 atoms of neon were present in 1 cubic centimetre, while during the night these figures were reduced to one tenth. These atoms are released by the radioactive disintegration of matter on the surface of the Moon or are conveyed to the Moon's surface by the solar wind.

3. The Moon's seismic activity is very low. Moonquakes on the lunar surface only reach the first or second point on the Richter scale and would not be noticed by man. These disturbances usually occur when the Moon is at perigee or at apogee. This has a direct effect upon the tidal forces of the Earth (i.e., the resulting interference with the Earth's gravity). Seismography can distinguish such ordinary lunar disturbances from the shocks caused by meteoric impact.

4. After the impact of a meteorite or a jettisoned rocket section the whole Moon starts to vibrate and these vibrations last several hours. This phenomenon is explained by the extensive dispersal of seismic waves through the rock formations in the crumbling lunar crust.

5. The average intensity of the magnetic field of the Moon is only 10^{-4} of that of the Earth's. However, in some areas magnetometers have measured an intensity ten to a hundred times greater. The cause of such anomalies has not yet been discovered.

6. The distribution of 'continent' and marial areas on the Moon is asymmetrical, as 'continent' formations predominate on the far side of the Moon.

7. Perturbations in the motion of man-made satellites orbiting the Moon led to the discovery of *mascons* (from mass-concentrations), which are concentrations of matter, usually located below the centre of mare (see p. 18). It is likely that these mascons, being local deposits of surplus matter, developed in those maria which were filled with lava, often 100 kilometres in diameter and 1 kilometre in depth. The dating of lunar rock samples brought back to Earth indicates that in origin these maria and their subsequent lava filling were distinct events possibly separated by several hundred million years. In fact the age of marial basalts is about 3 200 to 3 400 million years.

8. The results of seismic, gravitational, thermal, magnetic and other measurements have led to a more reliable understanding of the nature of the lunar interior. Regolith forms the surface *crust*, which is a crumbled layer about 60 kilometres deep. This crust rests on a firm *mantle* about 1 000 kilometres deep, whilst the centre is made up of a plastic *core*, the temperature of which is about 1 300°C.

9. The age of lunar rocks is approximately 4 700 million years, corresponding to the period when the oldest rock layer on the Moon was remelted for the last time, although our satellite was still a developing body and had yet to be completed by the accretion of meteoric debris in the course of its journey through space.

10. The composition of regolith is similar to basalt, although the samples brought differ substantially from any material from the Earth. They contain no water, have a low content of volatile elements (sodium, potassium, rubidium and lead), while some are relatively rich in uranium, thorium, possibly titanium and other ores. The marial and terra materials of the Moon are markedly different. The maria consist predominantly of basalt rock, while the terrae are largely formed from anorthosite rock. The latter is composed of lighter material, which contains double the amount of aluminium and half the amount of iron found in the darker marial areas.

11. No micro-organisms have been found on the Moon.

12. The latest information about the Moon has provided more definite answers to questions about its origin. Analysis of the samples showed first of all substantial differences in the chemical composition of the Earth and the Moon. This decisively confounds the hypothesis that the Moon originated by separation from the Earth. Similarly, the hypothesis of the simultaneous and adjacent development of the Moon and the Earth in a protoplanetary cloud around the Sun is now doubtful because, if this were the case, both bodies should have a similar chemical composition. More acceptable is the third possibility that the Moon originated far from the Earth and was later captured by it.

13. It has been demonstrated that man can live and work on the Moon, if protected by a space suit and that he can quickly adapt himself to lunar gravity, which is one-sixth of that of the Earth.

Finally the Moon holds the scientific key to the solution of problems relating to the origin and development of the solar system, largely because it has not been exposed to the erosive effects of the atmosphere, hydrosphere and biosphere. Whereas on Earth geologists can rarely find rocks older than about 3 500 million years, the samples of lunar rock contain information about events which happened some 4 500 million years ago. Some of the samples must have witnessed the formation of the Earth and seen the Sun when it was not the same type of star as it is today. If this sort of information can be analysed and interpreted, then an important step will have been made towards not only a fuller understanding of the past of the solar system, but also to an understanding of the present and future development of the Earth.

Lunar probes

Luna (USSR), Ranger, Surveyor (USA) and the Apollo space crafts (excluding artificial satellites orbiting the Moon).

Probe Date of reaching the Moon	Map	Result
Luna 2 13 September 1959	12	The first probe to reach the Moon.
Ranger 6 2 February 1964	35	Unsuccessful attempt — the target reached but television cameras not successfully activated.
Ranger 7 31 July 1964	42	The first close photographs of the Moon; 4 308 photographs transmitted; final photographs provided details down to 1 m and below.
Ranger 8 20 February 1965	35	7 137 photographs transmitted.
Ranger 9 24 March 1965	44	5 814 photographs transmitted, final ones from a height of 530 m and showing 25-cm details.
Luna 5 12 May 1965	42	First unsuccessful attempt at a soft landing.
Luna 7 7 October 1965	29	Unsuccessful attempt at a soft landing; speed of fall about 1 km/s.
Luna 8 6 December 1965	28	Unsuccessful attempt at a soft landing; descent speed of 15—20 m/s.
Luna 9 3 February 1966	28	The first soft landing; transmission of 4 panoramic photographs with details down to 1 mm in the foreground; the first information about the firmness of the surface layer.
Surveyor 1 2 June 1966	40	Soft landing; transmission of 11 240 photographs with a resolution of detail down to 0.5 mm; first colour photographs.
Surveyor 2 23 September 1966	32	Unsuccessful attempt at a soft landing.
Luna 13 24 December 1966	17	3 panoramic photographs; mechanical soil-sampler.
Surveyor 3 20 April 1967	42	6 326 photographs transmitted to Earth; position of probe was picked

Probe Date of reaching the Moon	Map	Result
		out by a photograph taken by Lunar Orbiter 3; later Apollo 12 expedition landed close to Surveyor 3 and some of its parts were subsequently brought back to Earth.
Surveyor 4 17 July 1967	33	Unsuccessful attempt.
Surveyor 5 11 September 1967	35	19 118 photographs transmitted to Earth; chemical analyses of the surface undertaken; first surface sampler tested.
Surveyor 6 10 November 1967	33	This probe took 14 500 photographs and was programmed to take off and land again 2.5 m away in order to make stereo-pictures; in all it transmitted some 29 952 photographs.
Surveyor 7 10 January 1968	64	The first soft landing in the mountainous continental area; 21 038 photographs; mechanical scoop excavated seven trenches up to 40 cm deep.
Apollo 11 20 July 1969	35	The first men on the Moon: Neil Armstrong and Edwin Aldrin; M. Collins left orbiting in the command module; first rock samples brought back to Earth; seismometer and laser reflector installed on the lunar surface.
Luna 15 21 July 1969	38	Technical examination of a new type of lunar probe.
Apollo 12 19 November 1969	42	Conrad and Bean on the Moon, Gordon in orbit; very exact landing, 180 m from Surveyor 3; the first station ALSEP.
Luna 16 21 September 1970	38	The first automatic collection of a lunar rock sample; 101 grams of the loose surface layer collected in a column sample that reached down to a depth of 35 cm below the surface were brought back to Earth.
Luna 17 17 November 1970	10	The first automatic mobile laboratory, Lunokhod 1, was transported to the Moon; the vehicle was at work

Probe Date of reaching the Moon	Map	Result
		for ten months and drove 10 540 m; detailed photographs were taken by four panoramic television cameras, and mechanical and chemical tests were carried out; French laser reflector employed.
Apollo 14 5 February 1971	42	Shepard and Mitchell on the Moon, Roosa in orbit; manual two-wheel truck was used to transport the samples and equipment; the second station ALSEP established.
Apollo 15 30 July 1971	22	Scott and Irwin on the Moon, Worden in orbit; electric vehicle with a television camera operated from Earth was used for the first time; geologically very interesting expeditions undertaken to the mare, mountain chains and Hadley's rille; drilling carried out to a depth of over 1.5 m; the third ALSEP station set up; photogrammetric mapping of the Moon.
Luna 18 11 September 1971	38	An attempted soft landing on uneven terrain was unsuccessful.
Luna 20 21 February 1971	38	Lunar rock samples taken in the mountainous region.
Apollo 16 21 April 1972	45	Young and Duke on the Moon, Mattingly in orbit; geological research in the mountainous region concerned with ray craters; technical equipment used similar to that of Apollo 15; the first astronomical observatory with UV camera and spectrograph established; the fourth ALSEP station set up.
Apollo 17 11 December 1972	25	Cernan and Schmitt on the Moon, Evans in orbit; the first scientist on the Moon, namely the geologist Schmitt; research carried out in the geologically recent region; technical equipment employed, the same as for Apollo 15; this was the last expedition of the Apollo programme.

Probe Date of reaching the Moon	Map	Result
Luna 21 15 January 1973	24	The second automatic mobile laboratory, Lunokhod 2, transported to the Moon; this equipment was supplemented by a magnetometer and astrophotometer.
Luna 23 6 November 1974	38	Unsuccessful attempt to take a sample of Moon rock, because apparatus was damaged during landing.
Luna 24 18 August 1976	38	Lunar rock sample taken in SE part of Mare Crisium. Column sample down to a depth of 200 cm.

Man-made lunar satellites and probes, which after orbiting the Moon returned to Earth

Name		Date of entry into lunar orbit	Dates of start of return journey and of arrival back on Earth
Luna 10	USSR	3 April 1966	
Orbiter 1	USA	14 August 1966	
Luna 11	USSR	28 August 1966	
Luna 12	USSR	26 October 1966	
Orbiter 2	USA	10 November 1966	
Orbiter 3	USA	8 February 1967	
Orbiter 4	USA	8 May 1967	
Explorer 35	USA	19 June 1967	
Orbiter 5	USA	5 August 1967	
Luna 14	USSR	10 April 1968	
Zond 5	USSR		15—21 September 1968
Zond 6	USSR		10—17 November 1968
Apollo 8	USA		21—27 December 1968
Apollo 10	USA		18—26 May 1969
Zond 7	USSR		8—14 August 1969
Zond 8	USSR		20—27 October 1970
Luna 18	USSR	7 September 1971	
Luna 19	USSR	3 October 1971	
Luna 22	USSR	2 June 1974	

Discovery of the planets

Planets were discovered for the first time many thousands of years ago. It was their complex motions among the fixed stars that first attracted human attention. However, it was not until the time of Nicholaus Copernicus, that the inhabitants of the Earth discovered that their world was also an ordinary planet, which along with the others was in orbit around the Sun.

Before the beginning of the seventeenth century knowledge about planets was almost non-existent. It was only when the telescope was invented that planets were discovered for a second time, and then they were identified as distant worlds similar to the Earth. Apart from the planets already known, another three were discovered, namely Uranus in 1781, Neptune in 1846 and Pluto in 1930. Recent technical developments have enabled a thorough investigation of the planets to take place, which in turn has led to their classification according to certain criteria. Firstly, according to whether the orbit lies inside or outside that of the Earth, the planets are divided into interior and exterior categories. Secondly, according to their size they can be divided into planets of a terrestrial type (e.g. Mercury, Venus, Earth, Mars, Pluto) and giant planets (e.g. Jupiter, Saturn, Uranus, Neptune).

Planet	Mean distance from the Sun (Mkm)	Sidereal period (in years)	Equatorial diameter (km)
Mercury	58	0.24	4 880
Venus	108	0.62	12 100
Earth	149.6	1.00	12 756
Mars	228	1.88	6 790
Jupiter	778	11.86	143 200
Saturn	1 427	29.46	119 300
Uranus	2 870	84.02	48 400
Neptune	4 497	164.78	49 200
Pluto	5 910	248.4	5 900

Relatively recent telescopic observations have shown that Venus and Mars most closely resemble Earth. As late as the 1950s scientists believed that life might be present on those planets, and of course science-fiction writers wrote continually about Martians and Venusians.

However, in the 1960s, when these planets were being rediscovered for the third time by space flights, it was demonstrated that the Earth was the only oasis of life in the solar system. The philosophical and practical repercussions of this discovery have not yet been fully appreciated.

Naturally the nearest planets were the first to be systematically studied. The exploration of Venus started with space probes in 1962 and the first space craft by-passed Mars in 1965. In December 1973 the first probe flew past Jupiter and in the spring of 1974 the first television shots of Mercury, which indicated a surprising similarity with the lunar surface, were the subject of public admiration. In December 1974 another probe approached Jupiter and exploited the acceleration provided by the gravitational field of this giant planet in order to continue its flight towards Saturn, which it will reach in September 1979. Given these ventures it is expected that the basic space research of the solar system will be completed during the 1980s.

VENUS

After the Sun and the Moon, Venus occupies one of the most prominent positions in the sky. Apart from the Moon it is the brightest celestial body and of all the planets approaches the nearest to the Earth, its distance from our planet being a mere 40 million kilometres, which is about a hundred times the distance of the Moon from the Earth. Venus can be seen with the naked eye even in the clear sky during daytime. Sometimes it shines long into the night, at other times it is visible before dawn, but never at midnight; at certain periods during the year it is not visible at all. Such an unusual behaviour certainly deserves some explanation.

The Evening and Morning Star

Figure 4 illustrates the motion of Venus round the Sun as seen from the Earth, or rather from a certain vantage point. The orbit of Venus is distinctly visible here and it is evident that the retreat of Venus on either side away from the Sun (to the east and to the west) is determined by the diameter of the planet's orbit.

From position 1 to position 5 Venus moves, when viewed from the Earth, to the left, that is to the east of the Sun and therefore it sets later than the Sun and thus appears in the sky as the *evening star*. In point 3 it is in the position farthest from the Sun and it is at this *greatest eastern elongation* that its angle from the Sun equals 47°.

After point 5 the planet passes to the right of the Sun and is not therefore visible in the evening. It then appears before sunrise and can be seen as the *morning star*. At point 7 its angle is 47° to the west of the Sun and at this point reaches its *greatest western elongation*. The periods when Venus is visible as the evening or morning star are shown in the following table:

Year	VENUS VISIBLE AS Evening Star	Morning Star
1976	beginning of July — end of the year (and longer)	beginning of the year — end of May
1977	beginning of the year — end of March	mid April — end of the year
1978	beginning of February — end of October	beginning of November — end of the year (and longer)
1979	beginning of October — end of the year (and longer)	beginning of the year — mid August
1980	beginning of the year — beginning of June	mid June — end of the year (and longer)
1981	beginning of May — end of the year	beginning of the year — beginning of March
1982	beginning of December — end of the year (and longer)	end of January — beginning of October
1983	beginning of the year — end of August	end of August — end of the year (and longer)
1984	same conditions as in 1976	etc.

This table can be applied to any point in time as the conditions for the visibility of Venus repeat after eight years. As an evening star Venus is at its highest point above the horizon at eastern elongations in springtime, and as a morning star it is usually best visible in autumn.

Venus Through the Telescope

The telescope was used for astronomical purposes for the first time in 1610 by Galileo Galilei, who discovered that Venus has *phases* similar to those of the Moon. When Venus is at superior conjunction (position 1 in Fig. 4), it can be seen almost as a complete disc, but because it is then at its greatest distance from the Earth, it appears as a small disc, a mere 10″ in diameter. Its angular diameter increases as the planet comes closer to the Earth. Venus reaches its greatest eastern elongation 221 days after its superior conjunction and the planet is visible as a *'quarter'*. It then continues its journey until after a further 71 days it reaches inferior conjunction as the *new Venus* (i.e. between the Sun and the Earth, position 5 in Fig 4). On rare occasions at inferior conjunction there is a *transit* of Venus when the planet passes across the solar disc. The last *transit* of Venus occurred on 6 December 1882 and the next will take place on 8 June 2004 and then on 6 June 2012.

Venus is brightest about 30 days before and after inferior conjunction, and 71 days after its inferior conjunction it is again visible as a 'quarter'. As Venus recedes from the Earth a position is reached where one half of the planetary disc is illuminated; this is the state of *dichotomy*, which is the moment when the disc of the planet is divided into two halves by the terminator. In another 221 days Venus is again at superior conjunction; the time lapse between two superior conjunctions is 583.92 days. This is called the *synodic period* of Venus, and as five synodic periods are equal to 8 years, similar aspects of Venus recur at intervals of 8 years.

Even large, astronomical telescopes have revealed little more than information about the phases of Venus. Only an experienced observer can distinguish on the disc or crescent

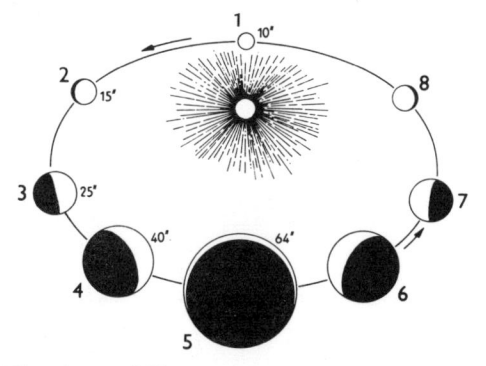

Fig. 4. *The phases of Venus*
The size of the disc indicates the angular diameter, which is also given in arc seconds. 1 — superior conjunction 5 — inferior conjunction 4, 6 — Venus at its maximum brilliance

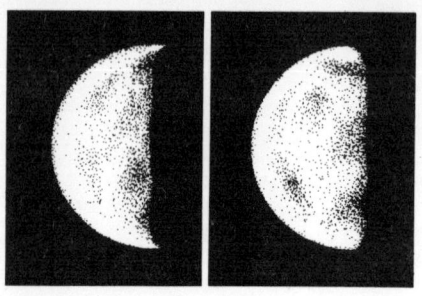

Fig. 5. *The appearance of Venus through the telescope*

of the planet some ill-defined, greyish shadings, which are very difficult to describe or express exactly in terms of position, shape and intensity. Figure 5 shows two typical examples of the appearance of Venus in a telescope, but here the intensity of these shadings is considerably stronger than in reality. Venus can best be observed during twilight or daylight, as its striking brightness in the darkness of the night sky extinguishes any minute details.

On looking through the telescope, it is not the surface of the planet which is visible, but the impenetrable layer of clouds. Because of this, Venus was until recently one of the greatest mysteries of the solar system, despite its proximity to Earth.

What We Know about Venus

Classical astronomy encountered a problem when it tried to determine the *rotational period* of Venus. Various estimates ranged from 24 hours to 225 days. It was only with the development of radioastronomy which took up where optical astronomy left off, that this period was finally determined at 243 days and that it is rotating in a retrograde direction (i.e. opposite to the direction in which the planet moves round the Sun). Venus rotates about an axis which is inclined at an angle of about 85° to its orbital plane.

Until recently it was not known with any certainty what was hidden under the Venusian cloud layer. However, two ways of penetrating to the planet's surface have been discovered, namely by radar and space probes.

A radiotelescope most commonly used to receive radio signals from space can also be used as a giant radiolocator to transmit signals to Venus. Not even the dense atmosphere is any obstacle for microwaves of 10 centimetres or less. The signals penetrate to the surface of the planet, are reflected and then return to Earth. They provide a range of information about the distance of the planet, its rotation, shape, nature of its surface etc.

Using radiolocation it has been discovered that the surface deviations of the planet from a true sphere are small. The solid body of Venus is a sphere of diameter 12 100 kilometres. The heights of features on the surface range between 3 and 5 kilometres. The radar mapping of Venus has supplied some very interesting results. The first radar maps appeared in 1964. Figure 6 shows one such map derived from radar observations by Haystack and Goldstone (USA) at Arecibo (Puerto Rico) in 1969. The light areas on the map correspond to intense radar reflectivity and may in fact be hilly regions, while the dark areas correspond to surfaces of low reflectivity (probably smoother surfaces). At the beginning of the 1970s, radar experiments suggested the presence of numerous craters on the planet's surface.

However, the most interesting information has been acquired by direct investigation of Venus using automatic probes, which are of three types: passing probes and man-made satellites, which measure the physical characteristics of the planet from a distance; and landing probes, which descend to the surface of Venus. The Soviet probes Venus (Venera) numbers 4 to 8, pioneered the work in this field. The probes weighing over 1 100 kilograms carried a shock-resistant ball (weight 400 kilograms), which before entering the atmosphere became detached and descended to the surface of the planet by parachute. The American Mariner 10 was the most extraordinary of the Venus probes, being the first space probe to transmit to Earth photographs revealing the nature of the cloud layer. The probes Venus 9 and 10 then made the first photographs of the stony surface of the planet.

The results of such direct investigation of Venus have considerably changed our concept of the planet, which used to be considered very similar to the Earth. Early optimistic beliefs in the existence of oceans and vegetation on Venus were definitely discounted when the space probes measured the temperature and pressure of the atmosphere. The temperature on the surface of Venus is about 470°C and the pressure about 9 MN/sq m. The density of the Venusian atmosphere is sixty times greater than that of water. There is practically no difference in the surface temperature between the day and night sides of the planet.

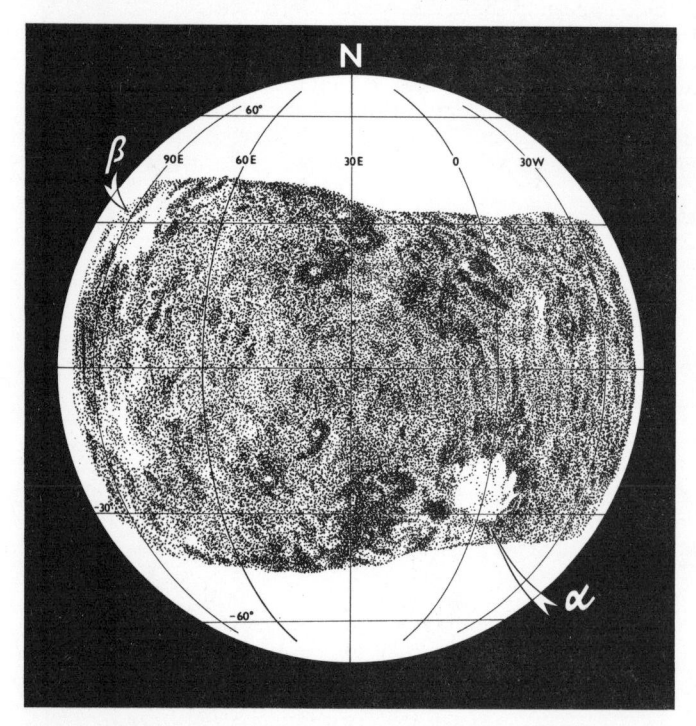

Fig. 6. *Radar map of a section of the Venusian surface*
The Greek letters alpha (α) and beta (β) indicate the surface areas which reflect radar signals most effectively.
α (0°, 25° S), β (85° E, 25° N)

The composition of the Venusian atmosphere is also completely different from the Earth's. It consists of about 95 per cent of carbon dioxide and 3 per cent of nitrogen, less than ½ per cent of oxygen, traces of water vapour and ammonia. The large amount of carbon dioxide in the atmosphere results from the high surface pressure and density. The high temperature of the planet can be best explained by the *greenhouse effect*. When the greatly filtered sunlight finally reaches the surface of Venus, it heats the surface which then gives off thermal radiation, which the atmosphere prevents from escaping. The resulting greenhouse effect produces only very slight fluctuations in temperature between day and night, which do not exceed 1°C.

The top of the opaque clouds is situated about 60 kilometres above the surface and is in turn covered by a turbid layer approximately 8 kilometres thick, which is composed of several further layers. The structure of the top layers of the clouds was first observed from ultraviolet photographs sent back by Mariner 10 (Fig. 7). Atmospheric thermals,

Fig. 7. *Venus as photographed by Mariner 10*

originating in the tropical zone, form long spiral strips in the upper cloud layer, which seem to encircle the planet. It has been confirmed that this cloud cover is perfectly continuous and that there is no gap in it through which the surface of Venus might be seen.

The clouds, however, are very transparent and the illumination on the surface of the planet is approximately the same as on the Earth when the sky is covered with a continous layer of clouds.

The most recent investigations of the planet have led to a number of very valuable discoveries, which in turn, as always happens in science, have given rise to a great many new questions. For example, why is the amount of water on Venus only 10^{-3} of that on Earth? What is the developmental stage of the planet's atmosphere? Why does Venus differ so much from other planets in the solar system? These and other questions can be answered only after a thorough investigation in the future, and to some extent it is left to the reader himself to try and complete this picture of the strange world of the planet Venus.

The Probes to Venus

The following table provides details of the Venus (Venera, USSR) and Mariner (USA) probe programmes. However, some experiments which did not give rise to any new information are not recorded.

PROBE Date of start	RESULT
Venus 1 12 February 1961	First attempted flight to Venus, passed the planet at a distance of 100 000 km on 20 May 1961. Communications interrupted 14 days after take-off.
Mariner 2 27 August 1962	First investigation of Venus. Passed at a height of 34 752 km above the planet on 14 December 1962. Measurements of the magnetic field, ionosphere etc.
Venus 2 12 November 1965	Passed at a height of 24 000 km above the planet on 27 February 1966. Communi-

PROBE / Date of start	RESULT
	cations interrupted before the planet was passed.
Venus 3 16 November 1965	First crash landing on Venus 1 March 1966. Transmission of results of measurements was unsuccessful.
Venus 4 12 June 1967	First detailed investigation of the atmosphere of Venus. Landing section entered the atmosphere in the night hemisphere on 18 October 1967 and descended by parachute. During the descent temperature, pressure, density and composition of the atmosphere were measured down to a height of about 26 km.
Mariner 5 14 June 1967	Passed at a height of 4 000 km on 19 October 1967. Seven types of measurements made (magnetic field, atmospheric pressure, temperature etc.). Measurement of radio reception behind the planet supplied valuable information about the atmosphere.
Venus 5 5 January 1969	Atmosphere of nocturnal hemisphere was entered on 16 June 1969. Landing section descended by parachute, and measurements were transmitted for 53 minutes until a height of about 25 km was reached. At this point the temperature was 320°C and the pressure 2.7 MN/sq m.
Venus 6 10 January 1969	Atmosphere of nocturnal hemisphere was entered on 17 May 1969, about 300 km away from Venus 5. Landing section transmitted measurements for 51 minutes during descent.
Venus 7 17 August 1970	First soft landing on Venus on nocturnal hemisphere on 15 December 1970. The apparatus transmitted data for some 23 minutes after landing. The recorded temperature on the planet's surface was 475°C and the pressure 9 MN/sq m. Chemical analysis of the atmosphere was carried out during descent.
Venus 8 27 March 1972	First soft landing in the diurnal hemisphere of Venus on 22 July 1972. Measurements of the chemical composition of the atmos-

PROBE Date of start	RESULT
	phere, wind velocity, light intensity, surface radioactivity of minerals, surface pressure and temperature. A surface temperature of 470°C and a pressure of 9 MN/sq m were recorded.
Mariner 10 3 November 1973	Passed Venus on 5 February 1974 at a height of 5 760 km. About 3 000 photographs were taken by two vidicon cameras with eight types of filters. These resulted in the discovery of the spiral structure of the cloud formations. Atmospheric circulation was also studied. Measurements of the chemical composition of the atmosphere and of temperature were taken but the extent of the magnetic field was not determined. The probe then passed on to Mercury and was the first to pass the planet on 29 March 1974 at a height of 750 km. 2 300 photographs were taken, which covered altogether 40 per cent of Mercury's surface. Mercury is similar to the Moon; it is densely covered with craters but does not have the larger 'seas'.
Venus 9 8 June 1975	First artificial satellite of Venus since 22 October 1975; on the same day landing section transmitted the first picture of the planet's surface, revealing numerous boulders. Measurements of parameters of the atmosphere and of the surface composition. Long-term observation of the planet from the satellite.
Venus 10 14 June 1975	Since 25 October 1975 second artificial satellite of Venus; on the same day landing section transmitted to Earth picture of another part of the planet's surface revealing flat terrain with numerous boulders. Details well visible up to the horizon, which is about 300 m distant.

MARS

Mars has attracted the astronomer's attention for a long time, particularly because of its reddish colouring, which is not characteristic of any other planet, and also for the extraordinary fluctuations in its brightness as well as its unusual motion among the stars. Legends and stories about the planet are well-known, as almost everyone has heard and read of the 'Martians' with their highly technical civilization, which is supposed to be very similar to that of the Earth.

Orbit Around the Sun

The distance between Mars and the Sun changes substantially, because the orbit of this planet is strongly elliptical. Its least distance from the Sun is 206 million kilometres when it is at *perihelion*, and its greatest distance from the Sun is 248 million kilometres when it is at *aphelion*. The orbital period of Mars is 687 days. The visibility of Mars depends on the relative positions of the Sun, the Earth and Mars. If the Earth is in a straight line with Mars and the Sun, Mars can be seen in the sky on the opposite side of the Sun; it is then at *opposition* and is visible all night. The distance between the Earth and Mars at opposition fluctuates between 55.5 million kilometres when Mars is at perihelion, and 101 million kilometres when Mars is at aphelion.

Observation of the planet is most favourable when Mars is in opposition at perihelion. Such oppositions occur every 15—16 years. The last such opposition occurred on 10 August 1971 and the next will take place on 28 September 1988. The period between two consecutive oppositions (the synodic period) fluctuates between 2 years 14 days and 2 years 80 days.

The red planet is smaller than the Earth, the diameter of Mars being 6 790 kilometres. On the other hand, Mars rotates on its axis in a very similar way to the Earth. The rotation of the planet was discovered as early as the seventeenth century by observation of its dark patches, which are visible even through small telescopes. A day on Mars lasts 24 hours 37 minutes 23 seconds of Earth time. Another similarity lies in the tilt of the planet's axis of rotation.

In consequence the inclination of the planet's equatorial plane to its orbital plane is 25°12'; the corresponding value for Earth is 23°30'. This tilt of the axis of rotation is responsible for the changes in seasons, which take place because the northern and southern hemispheres of the planet are not evenly irradiated by the Sun during the course of the year.

Mars Through the Telescope

Telescopic observation of Mars is not as easy as that of the Moon or of Venus. The angular diameter of the planet in the sky is negligible; its extreme values are shown in Fig. 8. When seen through a telescope, the disc of the planet at opposition resembles a lunar crater some 30—50 kilometres in diameter.

A beginner who looks through the telescope for the first time and expects to see views of the planet as he knows them from photographs and pictures in the literature will be disappointed. The details on the disc of Mars are very delicate and their visibility depends on the quality of the telescope, on the conditions for observation and also on the experience of the observer. The amount of detail which can be seen through the telescope depends on the aperture of the objective (assuming that this has good optical qualities). For example, an objective with an aperture of 120 mm should theoretically resolve two points which are 1″ of arc apart. Even the best telescopes in the world have a limit of resolving power of about 0.2″ — 0.1″.

A fundamental obstacle to telescopic observation is unsteadiness and blurring of the image arising from turbulence in the atmosphere, so it is necessary to wait for calm conditions. Patience is also needed to train the eye, improve observational techniques and draw an accurate picture of the details visible on the planet. A skilful amateur astronomer with good equipment can even now make valuable observations, providing that they are planned and systematic.

Three basic types of surface formations can be observed on the disc of the planet Mars and they are differentiated one from another by their *albedo* and colouring. The albedo is defined as the proportion of reflected light to the total amount of light received by the planet's surface. A bright

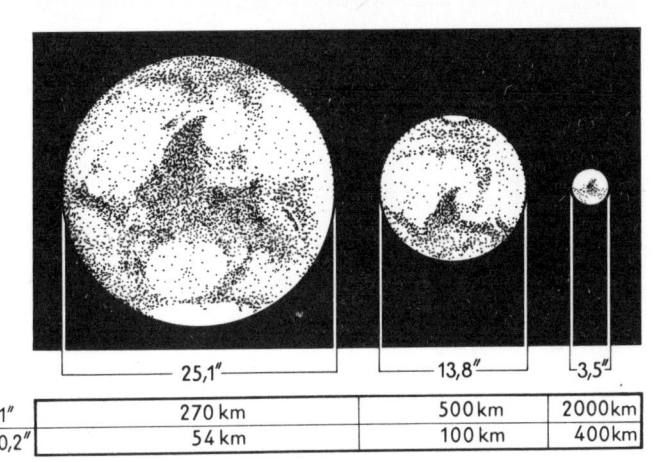

	25,1"	13,8"	3,5"
1"	270 km	500 km	2000 km
0,2"	54 km	100 km	400 km

Fig. 8. *Apparent diameter of Mars*
On the left at perihelic opposition, in the centre at aphelic opposition, on the right in conjunction with the Sun. At the bottom of the diagram diametres are given of visible features within a resolving power of 1" and 0.2".

surface has a high albedo and a dark surface has a low albedo.

The *polar caps* are the brightest phenomena; they are pure white and are instantly identifiable. The darkest areas on Mars are seen as greyish patches and are symbolically described as *seas (mare)*, *bays (sinus)* and *lakes (lacus)*. The remaining parts of the Martian surface are called *continents*. These cover some two-thirds of the planet's surface, are bright and have a striking reddish orange, salmon pink or brownish yellow colour. Such continents give Mars its characteristic colour, which distinguishes it in the starlit sky. The polar caps and the dark areas are the two characteristics of the planet that first attract the observer.

The visibility and shape of the polar caps depend on the season and on which Martian hemisphere is facing the Earth. The caps grow and diminish regularly according to the rhythm of the seasonal changes. The cap starts to increase in size in the autumn, when the associated polar region begins to disappear under a dense, whitish mist. During winter the cap continues to grow under this cloudy veil, but in spring the mist starts to disperse and reveals again the bright white polar cap, which is at this time at its largest and has a diameter of 50—70°, sometimes rising to even 100° (measured by calculating the angle from the centre of the planet).

Shortly after the mist has disappeared, the whole cap is visible and soon it starts to shrink and disintegrate into smaller sections, while its edges become lined with a dark rim. During summer the cap rapidly decreases, and finally only a small island remains. The remnant of the southern cap has an angular diameter of about 5° and its centre lies 6°30′ away from the South Pole. The northern cap survives the summer as a small island 6° in diameter with the centre 1° away from the North Pole.

The dark patches make up about one-third of the planet's surface, but although Mars looks similar to the Moon and its dark areas are also called seas, lakes etc., this does not mean that these formations are similar in substance. The Martian seas in fact constantly change their shape, size and shade of colour.

The most permanent, darkest, largest and also best visible features through a small telescope are the formations of Syrtis Major (see map D), Mare Sirenum (C), Sinus Sabaeus (A), Mare Tyrrhenum (D) and Mare Acidalium (A). Some maria have sharp outlines, while others merge imperceptibly with the light continents. Observations over a period of time have shown that this network of dark patches is continually changing and that these changes can be either illusory or real.

The *illusory changes* are caused by the obliteration of surface details by various types of clouds and mists that are present in the Martian atmosphere. Although cloud formations are not as abundant in the thin, dry and cold atmosphere of Mars as they are on Earth, they are a common and easily distinguishable phenomenon. Basically, three types of clouds are known: white, blue and yellow. *White clouds* form the dense cover above the winter polar region, but can also be found spasmodically in the equatorial zone and in the temperate regions. They can also take the form of a thin white mist or white mountainous clouds which gather above elevated areas, such as Nix Olympica. *Blue clouds* can be

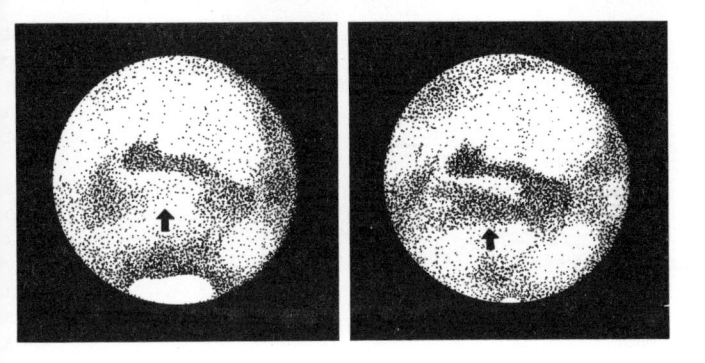

Fig. 9. *The appearance of Mars through the telescope*
Two drawings of different seasons showing periodic changes in the landscape Pandorae Fretum (see arrow).

found in similar areas to the white clouds, but are characterized by their reflection of blue and ultraviolet rays. *Yellow clouds* are the product of dust storms. According to their size they can be divided into the products of dust storms of planetary dimensions, smaller regional yellow clouds and a lingering yellow mist which remains after dust storms. These yellow clouds are often responsible for changing the usual appearance of the Martian surface beyond recognition. For example, at the end of September 1971, there was a global dust storm which was still in progress when the Mariner 9, Mars 2 and Mars 3 probes reached the planet, so that the planned photographic programme had to be postponed. Later, the Mariner 9 photographs recorded the passing of the storm and clearing of the atmosphere.

Real changes in features fall into two categories: accidental and seasonal. The characteristic patches change their size, albedo and colouring. A typical seasonal change is shown in Fig. 9; the picture on the left shows the situation at the end of spring in the southern hemisphere, when the region of Pandorae Fretum and the neighbouring Sinus Sabaeus regularly become darker. In the second half of autumn Pandorae Fretum lightens again. The seasonal changes are sometimes described as the 'dark wave', which spreads in spring from the pole to the equator. The speed at which the maria darken is about 30 kilometres per day. Before the era of interplanetary space flights these changes in albedo were often explained as changes in the humidity caused by the presence of vegetation, mosses and lichens, or were ascribed to Martian microorganisms. However, this conception of the Martian flora rightly belongs to the world of fantasy as do the reports on the mysterious 'canals' on Mars. These 'canals' were discovered by the Italian astronomer Schiaparelli (1877) and called by him *canali* (channels). They appear as dark narrow lines, often on the limit of visibility, which lead out from the maria or occasionally link marial or lake areas together. This theory caused a sensation when the American astronomer P. Lowell suggested that these formations were part of a complex irrigation system and were the product of an advanced technical civilization on the planet. Later, however, under ideal viewing conditions these canals were identified as a large number of small irregular patches which merge into continuous lines when insufficient detail is revealed.

The amount of guesswork and hypotheses which in the course of time became associated with Mars could only be finally resolved by direct investigation of the planet through the use of interplanetary probes.

Mapping of the Planet

In 1840 Beer and Mädler made the first attempts to transfer the details of individual sketches of Mars into a network of meridians and parallels. They also established the prime meridian, which passes through a dark patch close to the equator, at present known as Sinus Meridiani. The father of the modern mapping of Mars was G. Schiaparelli, who in 1877—79 worked out accurate maps based on micrometrical measurements. He also suggested a terminology for the various features, selecting terms from classical geography and mythology. He placed these names along the planet's equator so that a reading from east to west described the journey of the mythological sun chariot. Helios spent the first night in the Sun Lake (Solis Lacus), then during the red morning sky (Aurorae Sinus) he set out on his journey across the Indian Ocean (Mare Erythraeum) towards the lands of Indo-China (Chryse, Argyre), India (Margaritifer Sinus), Arabia (Thymiamata, Sabaeus Sinus), North Africa (Aeria, Syrtis Major, Lybia) and southern Europe (Hellas, Ausonia, Eridania), before arriving at the lands of the 'Hercules' columns' (Elysium, Atlantis, Amazonis), only to finally sink again in the far west into the Sun Lake.

Among the authors of the classical maps of Mars, the French astronomer E. M. Antoniadi must be mentioned, for his work from the years 1909—24 is still an example of accurate and technically perfect mapping. It was thanks to him that the sharp lines of the canals finally disappeared from the maps of Mars.

As the details of Mars appear different at each opposition, a different map has to be compiled for each one. Apart from this, there has also to be a general reference map, which shows only the relatively permanent formations and their names. To achieve uniformity, the IAU, in 1958, published an international map with 128 names and with the location of the named formations. However, some older traditional terms are still employed even today, and the meanings of the Latin names of the basic types of surface formations are as follows:

mare	— sea	*promontorium*	— cape
lacus	— lake	*fons*	— spring, beginning
palus	— marsh	*regio*	— landscape
sinus	— bay	*depressio*	— depression
fretum	— straits	*insula*	— island

The system of coordinates on Mars is similar to that of other planets. The basic great circle is the equator, whose position is determined by the position of the axis of rotation of the planet. Meridians pass through both poles at right angles to the equator. *Areographic coordinates* (i.e. the coordinates on Mars) consist of *areographic longitude L* and *areographic latitude B*. Longitude is measured from the prime meridian westwards from 0° to 360°. As Mars rotates, the meridians with an increasing longitude pass through the centre of its disc. Latitude is measured from the equator to both poles from 0° to 90°, positively to the north and negatively to the south. From 1961, on the recommendation of IAU, the maps of Mars have been orientated with the north at the top.

Astronautical projects have brought a major qualitative breakthrough in the study of Mars. All maps drawn according to observations from Earth are now called *albedo maps*, because they show only the areas having a different albedo. A new range of possibilities has been discovered since space probes began to transmit television pictures of the planet's relief to the Earth. These were the genesis of new specialized maps of Mars, such as morphological, geological and meteorological ones.

In 1973, the XV General Assembly of the IAU adopted a new system of nomenclature. The surface of the planet was divided into thirty sections, each named after the most prominent albedo feature. Craters are henceforth to be designated by a short word for the name of the section together with a combination of two to three letters according to a specific system. Apart from this, 189 large craters were given names of deceased naturalists, astronomers who studied Mars, physicists, geologists, biologists and also writers, painters and navigators. Out of these 189 names, 130 are also included in the nomenclature of the Moon.

Names were also given to the largest volcanoes (e.g. Mons Olympica) and a vast canyon area was named Valles Marineres to commemorate the Mariner project. The position of zero meridian is now defined by the position of a crater pit, Airy-0 (zero), which is 0.5 kilometres in diameter and lies inside a small crater, Airy, about 300 kilometres to the south of the equator.

A Close View of Mars

The validity and accuracy of the information about Mars accumulated by terrestrial astronomy was subjected to its severest test when confronted with the results of investigations made by the space probes. Some conclusions were confirmed, some hypotheses were refuted, but of prime importance was the completely new, unexpected and surprising evidence, which was so extensive that it considerably augmented the existing state of knowledge.

The first successful mission was that of the Mariner 4 probe in 1964, when the first photographs of Martian craters were transmitted to Earth. Initially Mars appeared very similar to the Moon, but this opinion was later modified by the photographs made by Mariners 6 and 7. Then a complete picture of the surface of Mars was provided in 1971—72 by the collection of 7 329 photographs transmitted by Mariner 9; these have been further supplemented by photographs taken by the Soviet probes to Mars. Apart from sending back photographs, these missions have also supplied data for measuring the physical characteristics of the planet and its atmosphere.

A general notion of the relief of the Martian surface is given in maps A—F (pp. 228—239). Geologically, Mars is divided into southern and northern hemispheres; the southern hemisphere consists of the old continental terrain densely covered with craters, which in many aspects resembles the lunar crater-covered continents. The northern hemisphere is more than 3 kilometres lower and is covered with basalt lava layers and dotted with a small number of craters. Basalt volcanic activity has also created sizeable structures, of which the largest is the volcanic peak Mons Olympica — a cone-shaped mountain which has a diameter of 500 kilometres at its base and is 24 kilometres high.

The relief of Mars offers a variety of shapes and forms. Apart from craters, there are other irregular terrains; some are covered by an irregular network of short mountain ranges and valleys, others are shapeless, structureless wastes, such as Hellas or Argyre I, lacking any sort of detail, as if they were covered by sand. Mars has also a variety of canyons; some resemble dry river beds, others mare depressions or crevices. The land surface is often wrinkled like an elephant's skin and also features a mountainous terrain with low, rounded mountain ranges similar to the wrinkle ridges on the Moon. Volcanoes, sedimentary layers, cracks and depressions are also found there.

The presence of an atmosphere, and in the past perhaps even of water, is manifested by a more advanced state of erosion than on the Moon. A considerable impact on the formation of the Martian landscape has been made by wind erosion, for the speed of wind close to the surface reaches 50—70 metres per second. Dust and sand, carried at such great speeds, can quickly change the surface of the planet.

Dark and light regions and their changes

On the Moon, the difference in the relief of bright continents and dark sea areas is noticeable at first sight. However, on Mars it is not always the case that the albedo can be closely associated with the physical relief of the landscape — that is, the dividing line between differently shaded terrains does not necessarily correspond to the boundary between bright and dark areas. In some cases the dark areas are smooth and craterless, whilst in other instances it is the bright areas that are smoother.

The nature of the changes in albedo has nevertheless been explained. When the dust storm ceases and the dust settles down, dark and bright 'tails' and very dark patches appear close to many craters. These result from the distribution of material by strong winds. These craters with their dark tails can always be found together in large groups and their location corresponds to the position of Martian seas, lakes etc.

Seasonal and long-term changes which can be observed from the Earth are also caused by the action of the wind, which carries material from one place to another. This fact finally destroyed the hypothesis that sought to explain these changes in albedo in terms of the presence of vegetation.

'Canals'

The typical Martian 'canals', well known from telescopic observations, have only exceptionally any counterpart in the physical relief of the planet. They are not crevices or canyons, mountain ranges or valleys, with the exception of the canyon Coprates, but they are products of the small resolution power of the telescope, together with lack of objectivity on the part of the observer. Photographs taken by space probes have revealed at least five types of real canals or rilles, which often resemble dry, meandering river beds. Some of them could be the result of the activity of running water, while others are of volcanic and tectonic origin.

Polar regions

The white covering of the polar caps consists of a thin layer of hoar frost. The remnants of these caps, which do not melt even in summer, probably consist of a thick layer of ice. The cloudy polar cap consists of crystals of frozen water and carbon dioxide.

Atmosphere

The atmospheric pressure on the Martian surface varies between 300—800 N/sq m, depending on the absolute height of a given region which is measured from the centre of the planet to a given point. This range of pressure corresponds to a variation in height of approximately 20 kilometres on the planet. The atmosphere consists predominantly of carbon dioxide, although traces of ozone, water vapour, nitrogen and argon have also been discovered. According to measurements taken by Mars 5, condensed water vapour probably forms a layer about 50 micrometres thick, while on Earth it would be 1 centimetre.

Strong wind storms develop in the atmosphere with a speed of up to 250 kilometres per hour. Yellow clouds, which result from such dust storms, usually reach a height of about 6—8 kilometres, but can rise up to 50 kilometres. When the storm abates, it takes about 60 days for the dust particles to settle.

Temperature

The average yearly temperature on Mars is about 50—60°C lower than on the Earth. It ranges between —20°C in the equatorial area to —80°C at the poles. The differences in temperature between day and night in the equatorial region are about 110°C. The highest summer temperature in the area of the equator is 20—30°C. The temperature of the polar caps is about —120°C. The dark areas are warmer than the light areas during the day. Hot places have been discovered in the night hemisphere with a temperature of about 10—25°C higher than the temperature of the surrounding region.

All the above-mentioned figures refer to the surface of the planet. As a result of the thin atmosphere these temperatures decrease very quickly as the distance from the surface increases, so that at 30 centimetres above the surface it is about 50° colder. The magnetic field of Mars, measured by the probes Mars 2, 3 and 5 is about 10^{-3} of the Earth's magnetic field.

Martians

The reader who critically assesses the above-mentioned facts, will certainly not share the optimistic ideas of science-fiction writers about the possibility of life on Mars. The Martians as well as other intelligent life forms on Mars belong to the world of fairy tales. However, a definite answer to questions about the presence of lower forms of life on this planet can only be given by a direct investigation by probes like Viking on the surface of the planet.

Satellites of Mars

In 1877 the American astronomer A. Hall discovered with the aid of a new 65-centimetre refractor at the Naval Observatory, Washington, the two moons of Mars. They were given the names of the companions of the God of War, namely Phobos (Fear) and Deimos (Horror).

Phobos completes an orbit at a distance of about 6,000 kilometres above the planet's surface once in 7 hours 39 minutes. It therefore circles the planet more than three times a day, rising in the west and setting in the east. From Mars its visibility is roughly equal to that of Venus from the Earth.

According to photographs taken by Mariner 9 it is a shapeless body with an approximate size of 27 by 19 kilometres. The longer axis of Phobos is always directed towards Mars. The surface of this moon is covered by craters, the largest or which has a diameter of 5.3 kilometres. Phobos reflects only 5 per cent of the sunlight which it receives and so it is the darkest known body in the solar system.

Deimos is in orbit at a distance of about 20 000 kilometres above the planet's surface in 30 hours 18 minutes. It is also a shapeless rock and its dimensions are about 15 by 11 kilometres. Its longer axis constantly points towards Mars.

The craters on the Martian moons are definitely of impact origin, and their large number indicates the great age of the moons. The moons have one hundred times more craters per unit area than does Mars. This fact supports the view that there has been a very pronounced erosion of the Martian surface, and it is also supported by the presence of atmosphere and other factors.

Probes to Mars

The following table records the results of the Mars (USSR) and Mariner and Viking (USA) probes. Some unsuccessful and unofficial attemps which did not provide any new information are not mentioned.

PROBE Date of start	RESULT
Mars 1 1 November 1962	First attempt to reach Mars. Contact lost after 21 March 1963 at a distance of 106 275 000 km from Earth.
Mariner 4 28 November 1964	First successful probe. On 15 July 1965 it passed around Mars at a distance of 9 846 km from the surface. 21 photographs with details of craters down to 3 km in diameter were transmitted to Earth. Fresh information obtained about the atmosphere; measured surface pressure of 400—700 N/sq m. No magnetic field was detected.
Mariner 6 24 February 1969	Flight past Mars on 31 July 1969 at a distance of 3 436 km from the surface. 50 photographs obtained before encounter and 26 photographs from the flight above the equator. Investigation of the atmosphere and the surface of Mars. Discovery of new types of terrain, both chaotic and featureless. Confirmation of low atmospheric pressure close to the surface.
Mariner 7 27 March 1969	Flight past Mars on 5 August 1969 at a distance of 3 200 km from the surface. 93 photographs obtained at this distance

PROBE Date of start	RESULT
	and 33 photographs taken during passage over the region of the south pole. Same equipment as on Mariner 6. Minimum temperature of south polar cap measured at −153°C. Photographs distinguished details down to 300 m.
Mars 2 19 May 1971	Second man-made satellite of Mars; put into orbit around the planet on 27 November 1971. Investigation of physical characteristics of the planet. The landing section separated from the probe and fell to the planet's surface.
Mars 3 28 May 1971	Third artificial satellite. Put into orbit on 2 December 1971. Landing section successfully separated from probe and first soft landing on planet made between the formations Phaethontis and Electris, 45° S, 158° W. Measurements taken during orbit.
Mariner 9 30 May 1971	First artificial satellite of Mars. Put into orbit on 14 November 1971. Before the transmitter was switched off on 27 October 1972 the probe made 698 orbits and transmitted to the Earth 7 329 photographs covering the whole surface of the planet, including photographs of Phobos and Deimos. Numerous measurements of the temperature of the planet's surface, of the pressure and composition of the atmosphere.
Mars 4 21 July 1973	Flight past the planet on 10 February 1974 at a height of 2 200 km. Technical failure made it impossible to bring the probe into orbit around Mars as planned.
Mars 5 25 July 1973	This artificial satellite was put into orbit on 12 February 1974 and transmitted until the beginning of March 1974. Photographs taken of the southern hemisphere; measurements of the pressure and composition of the atmosphere.
Mars 6 5 August 1973	Attempted soft landing on Mars. Separation of landing section from probe on 12 March 1974. Landing equipment transmitted the first direct measurements of the

PROBE	RESULT
Date of start	

planet's atmosphere via the probe. However, contact was interrupted directly above the planet's surface. Landing point was 24° S, 25° W in the region of Mare Erythraeum.

Mars 7
9 August 1973

Unsuccessful attempt at a soft landing. Landing section was detached on 9 March 1974, but because of a technical failure the section bypassed Mars at a distance of 1 300 km.

Viking 1
20 August 1975

Probes Viking were constructed as artificial satellites of Mars (Viking Orbiter) with detachable landing sections (Viking Lander). Viking Orbiter was brought into orbit on 19 June 1976. Successful soft landing of Viking Lander on 20 July 1976 in the region Xanthe, 48W, 22.5N. First photographs and measurements transmitted from the planet's surface. Biochemical analysis of samples collected by mechanical sample collector, tests to ascertain the presence of microorganisms and meteorological measurements were carried out.

Viking 2
10 September 1975

Viking Orbiter 2 was brought into orbit on 7 August 1976. Photographs of the planet taken during orbit. Soft landing of Viking Lander 2 on 4 September 1976 in the region Utopia, 226W, 47.5N, not far from the crater Mie.

Explanatory Notes to the Maps of the Moon

The plate section of this book contains two maps of the Moon, which are divided into eighty-two colour plates — a general six-part map and a detailed map of the Moon in seventy-six plates.

The general six-part map gives an idea of the distribution of seas and continents on the near and the far sides of the Moon. It marks the distribution of craters and also contains the basic nomenclature needed for an initial orientation. The individual maps delineate parts of the Moon as seen from the Earth (I), from the east (II), the far side (III), from the west (IV), from the north (V) and from the south (VI). The scale of the maps is 1 : 30 000 000 both at the equator and the poles. Maps I—IV follow Mercator's method, while maps V and VI are stereographic projections.

The detailed map of the Moon in seventy-six plates constitutes the main part of the book. It is a classical map using orthographic projection, in which the Moon is shown as it appears from the Earth at zero libration. Such illustrations are particularly appropriate for the observer on the Earth, because the outline of craters and other details is the same as seen through the telescope, and therefore comparison between the map and telescopic reality is easy. The libration zone, however, cannot be absolutely depicted in this way as a whole, because the outline of the map is formed by the meridians 90 E and 90 W. The scale of a detailed map is 1 : 3 700 000; this represents a diameter of the Moon's picture of 939.5 mm.

The system of coordinates has been dealt with on pp. 22 — 24. The parallels are drawn as straight lines, the meridians as ellipses. This system of coordinates can be used to incorporate new details whose coordinates are known (e.g. the landing places of further lunar probes).

A schematic map of the Moon on the front endpaper provides a key for rapid location of individual plates. The maps are numbered from 1 to 76 from west to east and from north to south. The numbers of adjacent maps are printed in red alongside every map. In the bottom margin there is a miniature map of the Moon, which pinpoints the position of the map section in question. The numbers of the maps are also contained in the index of named formations on pp. 245—251. Apart from a serial number every individual map carries the name of an important crater.

The surface of the Moon is drawn as it appears when illuminated from the east. To stress the plastic character of the terrain, the shadows cast by the slopes facing the west are marked. In agreement with telescopic observation, circular craters at the middle of the Moon are shown as circular and towards the edges of the disc they become elliptical. When measuring the diameter of a crater from the map, it is therefore necessary to measure the length of the major axis of this ellipse.

The nomenclature on the maps includes the names of the craters and also markings of associated formations using capital letters of the Latin alphabet and small letters of the Greek alphabet. The reason for this is that it is not possible to describe all lettered craters completely (e.g. Kepler A, Kepler B etc.) on a small scale map; therefore only the letters are marked (e.g. A, B). In order to determine immediately which name belongs to any given letter on a map, the German astronomers Beer and Mädler introduced the following rule: when a lettered formation is associated with a named formation, the letter is placed against that side of the formation which is nearest to the named formation. On maps 1—76, the centres of these lettered craters are marked by a black dot and the capital letter is then usually located on the line linking this point and the name of the adjacent crater to which the given letter relates. Black arrows pointing at the relevant name are used in places where confusion might occur.

The text accompanying the maps provides a brief description of the area, and a short note on all the interesting objects for amateur observation. This is followed by a list of named craters with short biographical data about the persons after whom the craters were named. The selenographic coordinates of craters are also given to the nearest 1°. All craters have their diameter marked in kilometres, as well as the maximum depth in metres for many of them — for example, Crater (68km/2 500m) means a crater with a diameter of 68 km and a depth of 2 500 m. As a result of the generosity of the American scientists D. W. Arthur and C. Wood, who supplied the author with the results of their latest measurements, it is possible to publish some very interesting data also about the depth of smaller craters which were only approximately known before. The smallest of these formations are good objects for amateur astronomers to test the viewing quality of their telescope.

Red print is used in relevant maps to mark the places of touchdown and landings of the lunar probes and space ships listed in the table on pp. 29—32.

Explanatory Notes to the Maps of Mars

The general map of both hemispheres of Mars, printed on the back endpaper, provides orientation in terms of the relative positions and names of the most prominent albedo formations, which can be recognized by experienced observers with telescopes having an aperture of about 15 cm. The second, more detailed map of Mars, is divided into six sections, marked A—F (pp. 228—239). The scale of the map is 1 : 60 000 000. The equatorial region is mapped on sections A, B, C and D following Mercator's method, while the polar regions (maps E and F) are shown in stereographic projection.

The six-part map of Mars describes two views of the planet: from the Earth and from a space probe. From the Earth only albedo formations can be observed. They are painted blue and their nomenclature, as recommended by the IAU, is printed in black. However, only the relatively stable albedo formations are illustrated, regardless of seasonal and other changes. A brown colour has been used to outline craters, canyons, volcanoes and other features of the physical relief of the planet, which have been revealed by space probes, especially by Mariner 9. The names of the most prominent craters, accepted in 1973 by the IAU at its assembly in Sydney, are printed in red. The index of terms for the maps of Mars is on pp. 252—255.

The text accompanying the six-part map contains a short description of the relief based on photographs of Mars taken from the space probes.

PLATES

MOON

1. The Central Part of the Near Side of the Moon

The prime meridian passes through the centre of the map; the left edge is formed by the line of longitude 50° west (50W), the right edge by the line of longitude 50° east (50 E). At the top lies the line of latitude 60° north (60N) and at the bottom of the map the line of latitude 60° south (60S). All the illustrated area is visible from the Earth.

This area is almost entirely covered by the lunar maria. Along the northern edge is situated Mare Frigoris (Sea of Cold), whilst in the west a considerable section of Oceanus Procellarum (Ocean of Storms) is located. In accordance with Mercator's projection, the shape of the craters and maria is not distorted and the true circular shape of Mare Imbrium (Sea of Showers or Rains), bordered by high mountain ranges, is clearly visible. The northern edge of the Sea of Rains adjoins the characteristic Sinus Iridum (Bay of Rainbows) and a darker crater, Plato. Along the southern edge, by the lunar Carpathians, is the prominent crater Copernicus, with rays radiating from it, and to the west of it two foci of bright rays, the craters Kepler and Aristarchus.

Three smaller maria, Mare Cognitum, Mare Humorum and Mare Nubium occupy the area to the south of the equator in the western hemisphere. Close to the middle of the near side of the Moon is a very prominent trio of craters, Ptolemaeus, Alphonsus and Arzachel. Three other small maria, Mare Vaporum, Sinus Aestuum and Sinus Medii are situated to the north of the equator.

To the east of the prime meridian lie the areas of Mare Serenitatis (Sea of Serenity), Mare Tranquillitatis (Sea of Tranquillity) and Mare Nectaris (Sea of Nectar). A part of Mare Fecunditatis (Sea of Fertility) also stretches into this region.

The following prominent craters, both individual and grouped, should be remembered for purposes of general orientation:

Aristoteles — Eudoxus, between Mare Frigoris and Mare Serenitatis,

Hercules — Atlas by the eastern edge of Mare Frigoris,

Theophilus — Cyrillus — Catharina by the western edge of Mare Nectaris,

Stöffler — Maurolycus,

Tycho — Longomontanus — Clavius — Maginus.

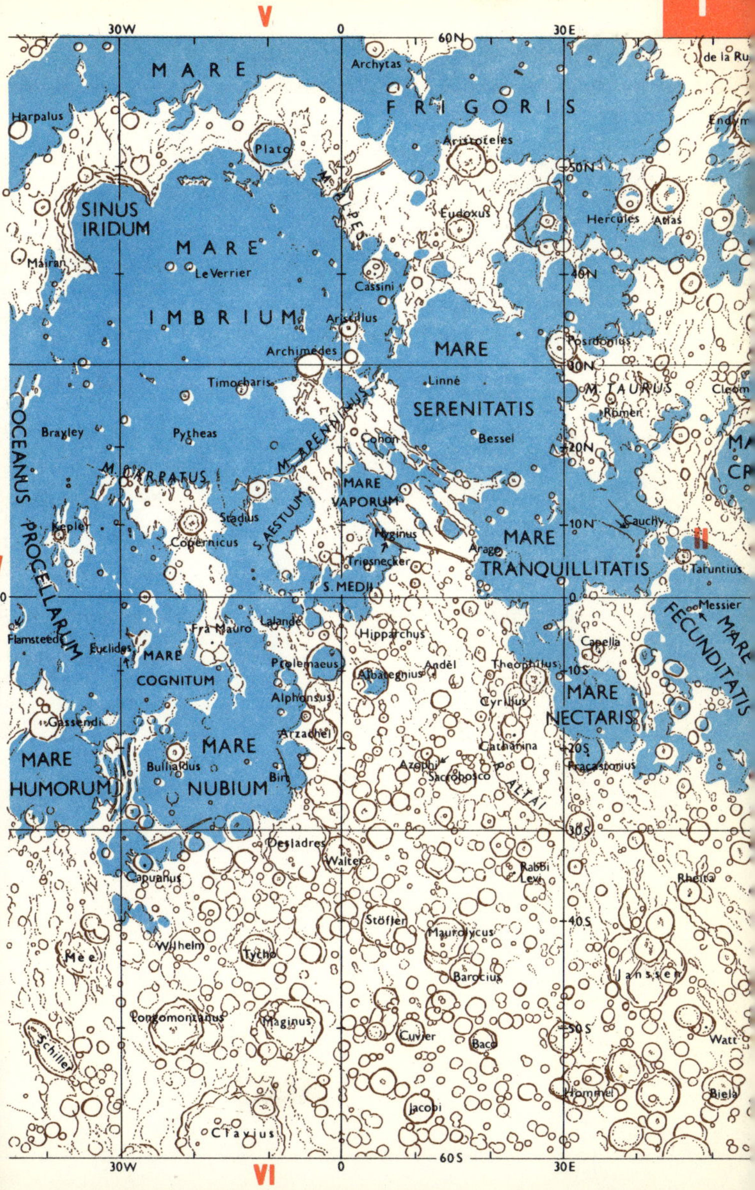
V
30W
0
60N
30E
MARE
FRIGORIS
Archytas
de la Ru
Harpalus
Endym
Plato
Aristoteles
M. ALPES
50N
SINUS IRIDUM
Eudoxus
Hercules
Atlas
MARE
Mairan
Le Verrier
Cassini
40N
IMBRIUM
Aristillus
Archimedes
MARE
Posidonius
30N
Timocharis
Linné
M. TAURUS
Cleom
SERENITATIS
Römer
OCEANUS
Brayley
Pytheas
M. APENNINUS
Conon
Bessel
20N
M. CARPATUS
MARE VAPORUM
PROCELLARUM
Stadius
AESTUUM
Kepler
Copernicus
S.
Hyginus
10N
Cauchy
MARE
II
Arago
Triesnecker
TRANQUILLITATIS
Taruntius
S. MEDII
0
Messier
Lalande
MARE
FECUNDITATIS
Flamsteed
Fra Mauro
Hipparchus
Capella
Euclides
MARE
COGNITUM
Ptolemaeus
Albategnius
Andel
Theophilus
10S
Alphonsus
Cyrillus
MARE NECTARIS
Gassendi
Arzachel
MARE
Catharina
20S
MARE
HUMORUM
Bullialdus
Azophi
Sacrobosco
Fracastorius
NUBIUM
Birt
R. ALTAI
30S
Desladres
Walter
Rabbi Levi
Capuanus
Rheita
Stöfler
40S
Maurolycus
Wilhelm
Mee
Tycho
Janssen
Barocius
Longomontanus
Maginus
50S
Schiller
Cuvier
Baco
Watt
Hommel
Biela
Jacobi
Clavius
60S
30W
0
30E
VI

II. The Central Part of the Eastern Hemisphere of the Moon

The middle of the map is intersected by the line of longitude 90E, which forms the eastern edge of the Moon when seen from the Earth during zero libration. In effect, because of libration, this border fluctuates about 7° on either side of 90E between the near and the far side of the Moon.

This is a side view of the Moon, similar to that obtained by the probe Luna 3, which in 1959 took the first photographs of the far side of the Moon.

Two dark patches along the northern edge of the map are the crater Endymion and Mare Humboldtianum. In the middle of the section is the circular Mare Smythii and further to the north stretches the elongated Mare Marginis with its crater Neper. Both these maria are visible from the Earth, but they are notably distorted, as they are positioned along the eastern edge of the Moon.

Along the western margin of the map the distinctive Mare Crisium is located in its true shape, and to the south of the equator the long Mare Fecunditatis, in the vicinity of which is a famous group of four craters, Langrenus — Vendelinus — Petavius — Furnerius.

In the southern part of the map Mare Australe (Southern Sea) stands out, broken up by a number of craters and light islands. During a favourable libration the flooded craters Lyot and Oken can be clearly seen from the Earth, running along the edge of the Moon as pronounced, elongated, dark ellipses. Mare Australe extends far into the rear side of the Moon.

The dark surface of maria on the far side of the Moon are rare and therefore orientation is more difficult. The best points of orientation are large and prominent craters, such as Compton (in the north), Joliot and Lomonosov (with a dark floor), Pasteur and Hilbert (a prominent pair) etc.

An excellent point of orientation on the far side is provided by the crater Tsiolkovsky, which has an unusually dark floor and a bright central peak.

II

- de la Rue
- Endymion
- Atlas
- MARE HUMBOLDTIANUM
- Compton
- Mercurius
- Messala
- Fabry
- Millikan
- H.G.Wells
- Kur...
- Gauss
- Cleomedes
- Plutarch
- Joliot
- Lomonosov
- Seyfert
- Vernadsky
- MARE CRISIUM
- MARE MARGINIS
- Fleming
- Neper
- Taruntius
- MARE SMYTHII
- Mende
- Messier
- Bečvář
- MARE FECUNDITATIS
- Langrenus
- Ansgarius
- Pasteur
- Vendelinus
- Sklodowska
- Hilbert
- Fermi
- Petavius
- Curie
- Tsiolkovski
- Humboldt
- Pa...
- Milne
- Rheita
- Furnerius
- MARE
- Roche
- Oken
- Lamb
- Pauli
- ...ssen
- AUSTRALE
- Lebedev
- Watt
- Lyot
- Biela
- Priestly
- Pontécoulant
- Planck

V — 60E / 90E / 120E / 60N
VI — 60E / 90E / 60S / 120E

I, III

III. The Central Part of the Far Side of the Moon

This map offers the view of the far side of the Moon as it would be seen from 'behind', i.e. against the Earth. The centre of the map is given by the 180° meridian and therefore the centre of the map is also the centre of the far side of the Moon. When this map is compared with map I, the surprising difference between the near and the far side in terms of the distribution of maria is immediately apparent.

The most prominent sea on the far side is Mare Moscoviense (Moscow Sea), which in size equals Mare Humorum or Nectaris on the near side. Mare Moscoviense was discovered on the first photograph of Luna 3 in 1959 together with a striking dark patch, namely the flooded crater Jules Verne, and an indistinct darkening along the margin of the photograph, later to be called Mare Ingenii (Sea of Ingenuity). Later photographs taken by the Lunar Orbiter satellites revealed that the expected large mare was only an area of flooded craters and basins and so the name of Mare Ingenii was given to one of these formations.

Orientation in the southern part of this area is relatively easy because of the presence of such dark craters as Jules Verne, Poincaré, Leibnitz, etc. Particularly striking are the basins Apollo and Korolev.

The area between Mare Ingenii and Mare Moscoviense includes numerous large and prominent formations; the craters Mendeleev and Gagarin are about 280 kilometres in diameter and the twin pair Keeler and Heaviside is a particularly distinctive feature.

The central part of the far side of Moon lacks such prominent formations, apart from the pair of craters Daedalus and Icarus. To the north of the centre there are indications of a wall of a large basin; its existence is also supported by a mascon discovered there. Apart from this the basin is well disguised by a number of more recent craters and the absence of dark, marial material.

The northern hemisphere in this section is covered by a group of large craters, D'Alembert, Campbell, Wiener and Kurchatov, and a twin pair, Birkhoff — Rowland, which are useful for establishing points of reference.

III

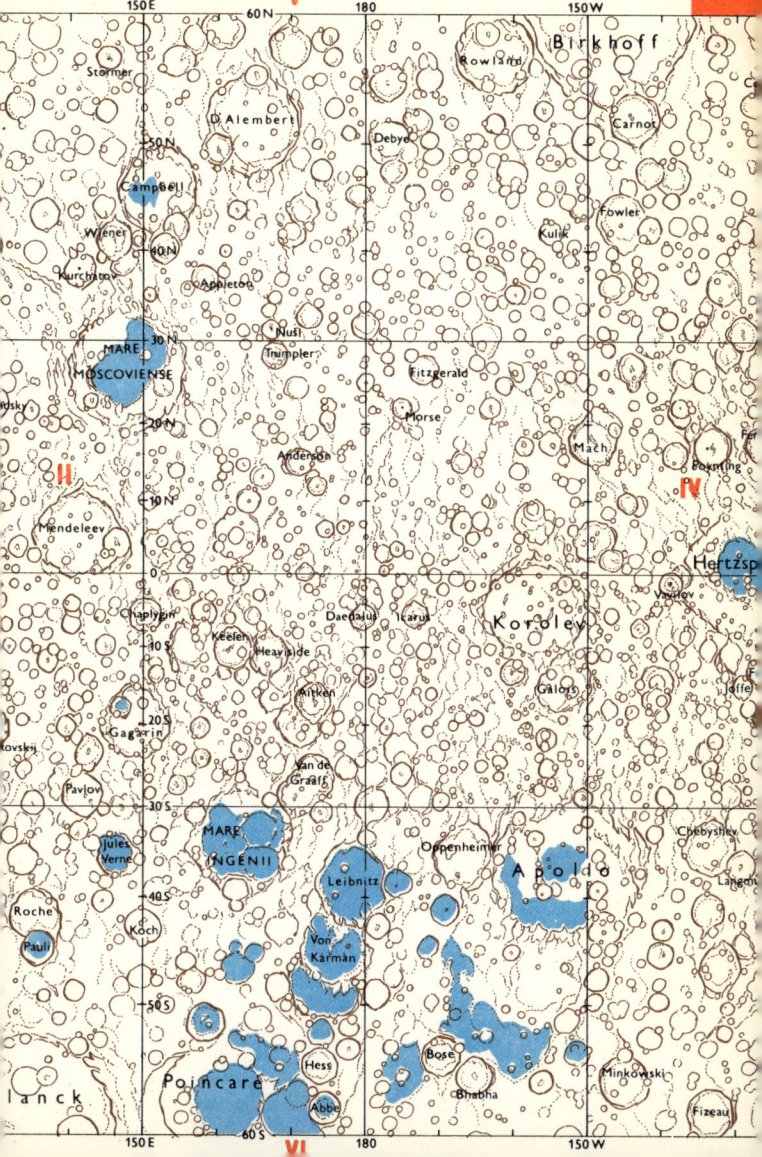

150 E | 60 N | 180 | 150 W

- Birkhoff
- Störmer
- Rowland
- D'Alembert
- Carnot
- Debye
- 50 N
- Campbell
- Fowler
- Wiener
- 40 N
- Kulik
- Kurchatov
- Appleton
- 30 N
- Nusl
- MARE MOSCOVIENSE
- Trumpler
- Fitzgerald
- 20 N
- Morse
- dsky
- Anderson
- Mach
- Poynting
- 10 N
- Mendeleev
- 0
- Hertzsp
- Chaplygin
- Daedalus
- Icarus
- Korolev
- 10 S
- Keeler
- Heaviside
- Vavilov
- Aitken
- Galois
- offe
- 20 S
- Gagarin
- ovskij
- Pavlov
- Van de Graaff
- 30 S
- Jules Verne
- MARE INGENII
- Oppenheimer
- Apollo
- Chebyshev
- Roche
- 40 S
- Leibnitz
- Pauli
- Koch
- Langm
- Von Kármán
- 50 S
- Hess
- Bose
- Poincaré
- Minkowski
- Abbe
- Bhabha
- lanck
- Fizeau
- 60 S

150 E | 180 | 150 W

IV. The Central Part of the Western Hemisphere of the Moon

The 90° meridian runs down the centre of the map and provides the ideal borderline between the near and far side of the Moon. The real borderline along the edge of the Moon, however, changes according to the way the Moon is turned during librations towards the Earth.

The view of the Moon from the west is very interesting because of the vast lunar basin Mare Orientale. The dark surface of Mare Orientale (Eastern Sea) lies between 90W and 100W, so that it is entirely located on the far side of the Moon and only during a very favourable libration does it reach the edge of the Moon, when it is visible from the Earth. Similarly, the sizeable mountain rings which surround the basin (i.e. Montes Rook and Montes Cordillera) can only be seen partially at the edge of the Moon, and then they are considerably distorted. The centre of the basin is the focal point of long chains of craters, valleys and rilles. These are several hundred kilometres in length and extend predominantly in north-west and south-east directions. The whole area of the basin is, unfortunately, not observable from the Earth.

In the right half of the map the familiar formations of the near side of the Moon can be observed. Firstly there is the dark area of the largest lunar mare, Oceanus Procellarum (Ocean of Storms), which in the north continues as Sinus Roris (Bay of Dew). The brightest formation in the Ocean of Storms is the crater Aristarchus. The pair of craters Struve — Eddington can be observed easily from the Earth close to the western edge of the Moon.

Close to the equator a conspicuous dark patch can be seen, namely the crater Grimaldi and the adjacent crater Riccioli. Another prominent group is formed by the three craters Schickard, Wargentin and Phocylides.

The border between the near and the far side is the location of the crater Einstein, which can be seen in its entirety from the Earth only on exceptional occasions. The most noticeable formation on the far side is the basin Hertzsprung with its flooded centre. There are also some large craters situated to the north of the equator, of which the largest are Lorentz and Landau. Finally, the southern hemisphere is characterized by the three craters Chebyshev, Langmuir and Brouwer.

IV

Map Section

Craters and features (north to south, approximate):

- Coulomb
- Xenophanes
- South
- Cannizzaro
- Volta
- Markov
- SINUS RORIS
- Stefan
- Galvani
- Landau
- Rümker
- Lorentz
- Röntgen
- OCEANUS PROCELLARUM
- Aristarchus
- Herodotus
- Struve
- Eddington
- Persman
- Robertson
- Ohm
- Einstein
- Marius
- Pointing
- Galilaei
- Michelson
- Hertzsprung
- Riccioli
- Grimaldi
- Fridman
- MONTES CORDILLERA
- Joffe
- MONTES ROOK
- MARE ORIENTALE
- Darwin
- Mersenius
- Vieta
- Chebyshev
- Brouwer
- Steklov
- Lagrange
- Langmuir
- Piazzi
- Baade
- Schickard
- Mendel
- Inghirami
- Arrhenius
- Wargentin
- Phocylides
- Fizeau
- Pingré

Grid labels: 120W, 90W, 60W; 60N, 50N, 40N, 30N, 20N, 10N, 0, 10S, 20S, 30S, 40S, 50S, 60S

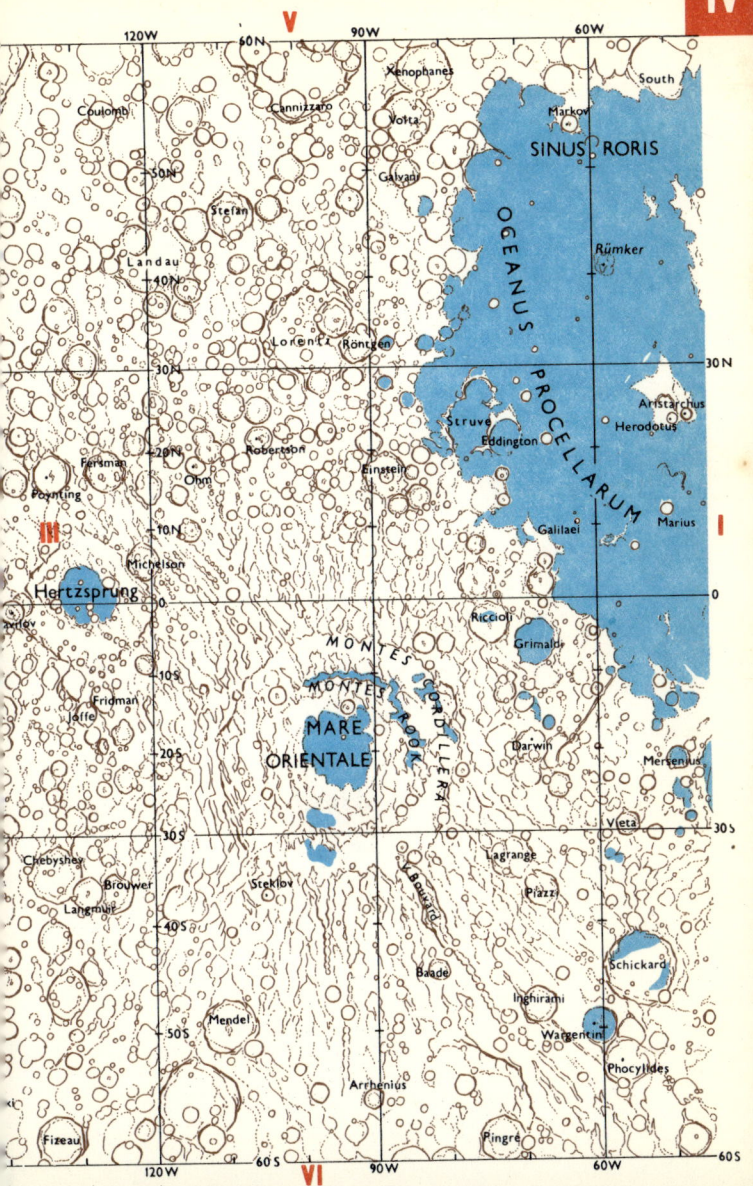

Section markers: V (top), VI (bottom), I (right), III (left)

V. The Northern Polar Region of the Moon

The centre of the map is given by the North Pole, which is intersected by the lines of longitude 90E and 90W and this is therefore the ideal dividing line between the near and the far side of the Moon. When the Moon is looked at from the North Pole along the line of its axis of rotation, a part of the near side is visible at the bottom, and a part of the far side as far as selenographic latitude 50 N on the upper part of the map.

Mare Frigoris, which is the northernmost mare on the Moon, is situated along the bottom edge of the map (c.f. map I). Higher up, in the direction of the North Pole, the walled plain Meton is situated. The area surrounding the North Pole outside the latitude 80N can be observed from the Earth with great difficulty, and such visibility is limited to the incidence of favourable librations. The crater Peary is situated closest to the pole and its southern wall is linked with the crater Byrd.

The right (eastern) edge of the map contains two dark areas, the crater Endymion with its flooded floor and Mare Humboldtianum, which extends to longitude 90E. The area extending beyond the Sea of Humboldt to the far side of the Moon contains two craters, Compton and Schwarzschild, each with a diameter of about 200 kilometres. Along the upper edge of the map the walled plain D'Alembert is situated, which stretches towards the centre of the far side of the Moon (c.f. map III). The largest formation in the western part of this polar region of the far side is the lunar basin Birkhoff, the diameter of which is about 300 kilometres.

Orientation near longitude 90W, which forms the western limb of the visible side of the Moon, is quite difficult, because of the dense crater field. However, there are very large craters, such as Xenophanes and Pascal, that can be relied upon as a last resort. The crater Brianchon stretches to the far side and is therefore difficult to observe.

To the south, not far from Sinus Roris, lies the very prominent crater Pythagoras, which has a large terrace-like wall and a central mountain chain. When viewed from the Earth, Pythagoras appears very close to the limb of the Moon, but can be seen during any lunation.

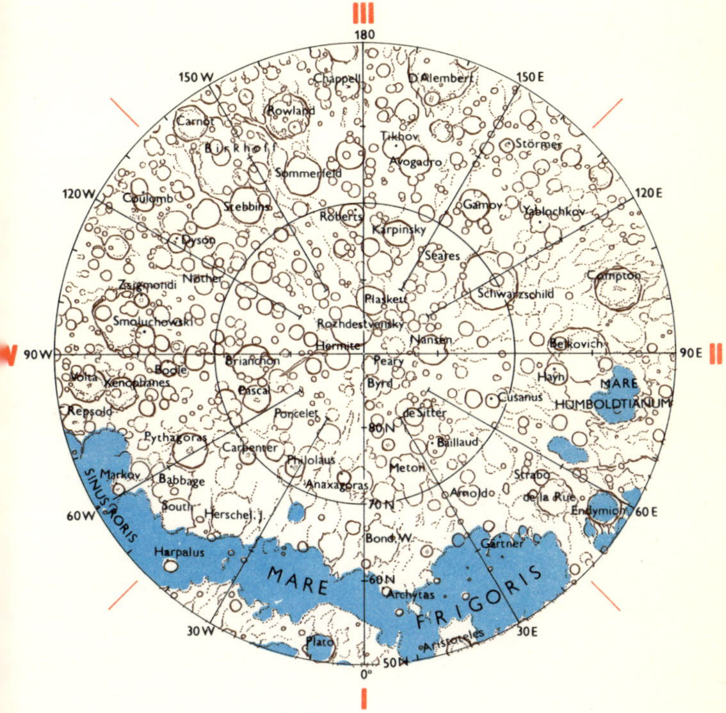
III
180
150 W
150 E
120 W
120 E
90 W
90 E
II
60 W
60 E
30 W
30 E
0°
I
80 N
70 N
60 N
50 N
Chappell
D'Alembert
Rowland
Carnot
Tikhov
Störmer
Birkhoff
Avogadro
Sommerfeld
Coulomb
Stebbins
Roberts
Gamow
Yablochkov
Karpinsky
Dyson
Seares
Compton
Nöther
Zsigmondi
Plaskett
Schwarzschild
Smoluchowski
Rozhdestvensky
Nansen
Hermite
Belkovich
Brianchon
Peary
Boole
Volta
Xenophanes
Hayn
Byrd
MARE
HUMBOLDTIANUM
Pascal
Cusanus
Repsold
Poncelet
de Sitter
Pythagoras
Baillaud
Carpenter
Philolaus
Meton
Strabo
Markov
Babbage
Anaxagoras
Arnoldo
de la Rue
SINUS RORIS
South
Herschel, J.
Endymion
Bond, W.
Gärtner
Harpalus
MARE
FRIGORIS
Archytas
Plato
Aristoteles

VI. The Southern Polar Region of the Moon

The South Pole is situated in the centre of the map, intersected horizontally by the ideal borderline between the near and the far side of the Moon. This is formed by longitude 90W (on the left) and 90E (on the right). The prime meridian rises upwards from the South Pole and the 180° meridian extends downwards to the centre of the far side of the Moon. The boundary of the map is given by latitude 50S.

The characteristic of this region is a uniform and compact terrain, dotted with craters and covered here and there with small areas of dark mare material. The only mare which partly stretches into this area is Mare Australe (Southern Sea), situated on both sides of longitude 90E.

On the near side of the Moon, in the area close to the prime meridian, are the craters Clavius and Moretus, which can be observed easily from the Earth. The visibility of the large walled plain Bailly is not so good, and it can be seen only as a narrow ellipse close to the southern edge of the Moon. The circular mountain chains Hausen and Drygalski can be seen only very occasionally.

The area adjacent to the South Pole is the most difficult part of the Moon to see, not only from the Earth but also from artificial lunar satellites. The terrain is very mountainous and because the Sun's rays illuminate the South Pole only at a very acute angle, the low-lying areas are almost permanently enveloped in shadow. Although photographs made by the probes Lunar Orbiter 4 and 5 helped to reveal a great deal of detail, this area around the South Pole and up to 100W and 120W is still largely a blank space on lunar maps. However, this unmapped area constitutes less than 1 per cent of the lunar surface.

The craters Amundsen and Scott are appropriately located close to the South Pole, but the wall of the crater Amundsen is visible only in exceptional circumstances.

On the far side of the Moon the vast basin Schrödinger is located, from which long valleys radiate. The basin Planck and Poincaré are a little smaller. Orientation in this area is made easier by the presence of three large craters Minnaert, Antoniadi and Numerov as well as the crater Zeeman.

A glance at the south polar region makes one appreciate the extent of progress made in mapping the Moon in the 1960s. Before 1959, the whole of the far side was practically bare, but by 1967 this 'bare hemisphere' had shrunk to a negligible area adjacent to the South Pole.

VI

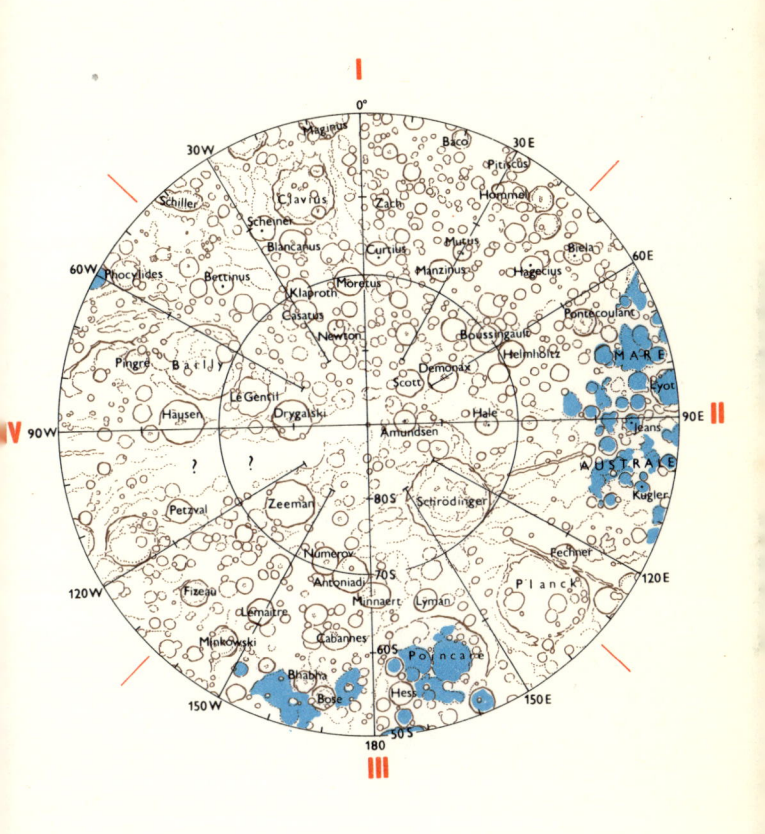

1 MARKOV

The map shows a section of the north-western portion of the near side of the Moon including Sinus Roris. The craters near the limb in the libration zone of the Moon are best visible shortly before Full Moon.

CLEOSTRATUS 76W, 60N) About 500 BC; Greek philosopher and astronomer; improved Greek calendar. ○ *crater (63 km)*

GALVANI (84W, 50N) Luigi Galvani, 1737–98; Italian physicist and physician; specialist in comparative biology.
○ *crater (75 km)*

LANGLEY (87W, 52N) Samuel P. Langley, 1834–1906; American astronomer and physicist; determined transparency of atmosphere for different wavelengths in the solar spectrum.
○ *crater (40 km)*

MARKOV (63W, 53N) 1. Andrei A. Markov, 1856–1922; Russian mathematician; specialist in the theory of probability. 2. Alexander V. Markov, 1897–1968; Soviet astrophysicist; photometry of the Moon. ○ *crater with a sharp rim (41 km)*

OENOPIDES (64W, 57N) from Chios, about 500–430 BC; Greek astronomer and geometer; discovery of the inclination of the ecliptic to the celestial equator is ascribed to him.
○ *walled plain (69 km)*

RÉGNAULT (88W, 54N) Henri Victor Régnault, 1810–78; French chemist and physicist; noted for work on physical properties of gases. ○ *crater (50 km)*

REPSOLD (77W, 51N) Johann G. Repsold, 1771–1830; German manufacturer of optical, mechanical and astrometrical precision equipment. ○ *disintegrated crater (107 km)*

RORIS, SINUS Bay of Dew. Mare area which links Mare Frigoris and Oceanus Procellarum; named by Riccioli.

STOKES (88W, 52N) Sir George G. Stokes, 1819–1903; British mathematician and physicist; laid foundations of hydrodynamics and spectral analysis; studied the figure and gravitational field of the Earth. ○ *crater (50 km)*

VOLTA (84W, 54N) Count Allessandro G.A.A. Volta, 1745–1827; Italian physicist; developed first electric battery in 1799.
○ *crater (109 km)*

XENOPHANES (80W, 57N) From Colophon, about 570–478 BC; Ionian philosopher, satirist and poet; believed in flat Earth theory. ○ *crater (111 km)*

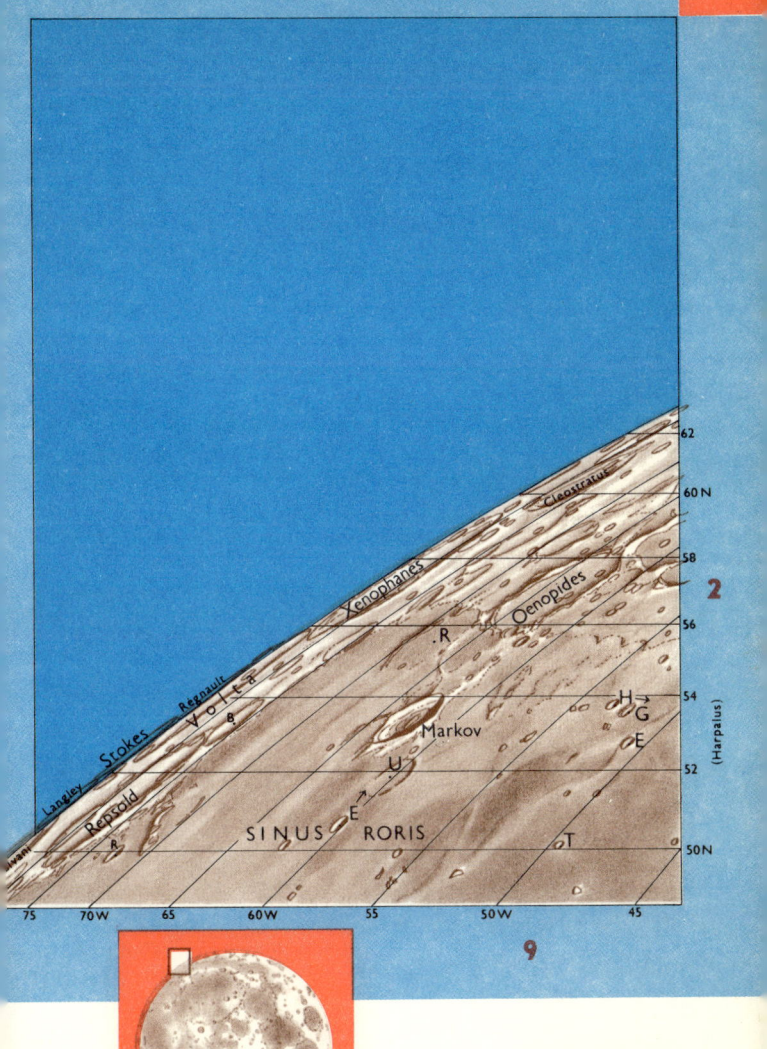

MARKOV

2 PYTHAGORAS

In this section of the Moon close to the northern margin of the visible disc, Sinus Roris borders on Mare Frigoris.

ANAXIMANDER (51W, 67N) From Miletus, about 611−547 BC; Greek philosopher; asserted that the Earth was cylindrical.
○ *crater (68 km)*

BABBAGE (57W, 60N) Charles Babbage, 1792−1871; English inventor of a calculating machine.
○ *walled plain (144 km)*

BIANCHINI (34W, 49N) Francesco Bianchini, 1662−1729; Italian astronomer. ○ *crater (39 km)*

BOOLE (87W, 64N) George Boole, 1815−64; English mathematician and logician. ○ *crater (57 km)*

BOUGUER (36W, 52N) Pierre Bouguer, 1698−1758; French hydrographer, geodesist and astronomer. ○ *crater (23 km)*

CARPENTER (51W, 69N) James Carpenter, 1840−99; English astronomer. ○ *crater (60 km)*

LA CONDAMINE (28W, 53N) Charles M. de la Condamine, 1704−74; French physicist, astronomer. ○ *crater (37 km)*

CREMONA (90W, 67N) Luigi Cremona, 1830−1903; Italian mathematician. ○ *crater (87 km)*

DESARGUES (74W, 70N) Gerard Desargues, 1593−1662; French mathematician and engineer. ○ *crater (83 km)*

FOUCAULT (40W, 50N) Léon Foucault, 1819−68; French physician and physicist; first demonstrated the nature of the rotation of the Earth on its axis. ○ *crater (24 km)*

HARPALUS (43W, 53N) About 460 BC; Greek astronomer.
○ *ray crater (40 km)*

HERSCHEL, J. (41W, 62N) John Herschel, 1792−1871; English astronomer; son of William Herschel.
○ *disintegrated walled plain (156 km)*

HORREBOW (41W, 59N) Peder Horrebow, 1679−1764; Danish mathematician and physicist. ○ *crater (25 km)*

MAUPERTUIS (27W, 50N) Pierre Louis de Maupertuis, 1698−1759; French mathematician and geodesist.
○ *disintegrated crater (46 km)*

PYTHAGORAS (62W, 63N) About 500 BC; mathematician and astronomer; founder of Greek school of philosophy and science.
○ *very prominent crater (128 km)*

ROBINSON (46W, 59N) John T. R. Robinson, 1792−1882; Irish astronomer and physicist. ○ *crater (24 km)*

SOUTH (50W, 57N) James South, 1785−1867; English astronomer. ○ *disintegrated walled plain (98 km)*

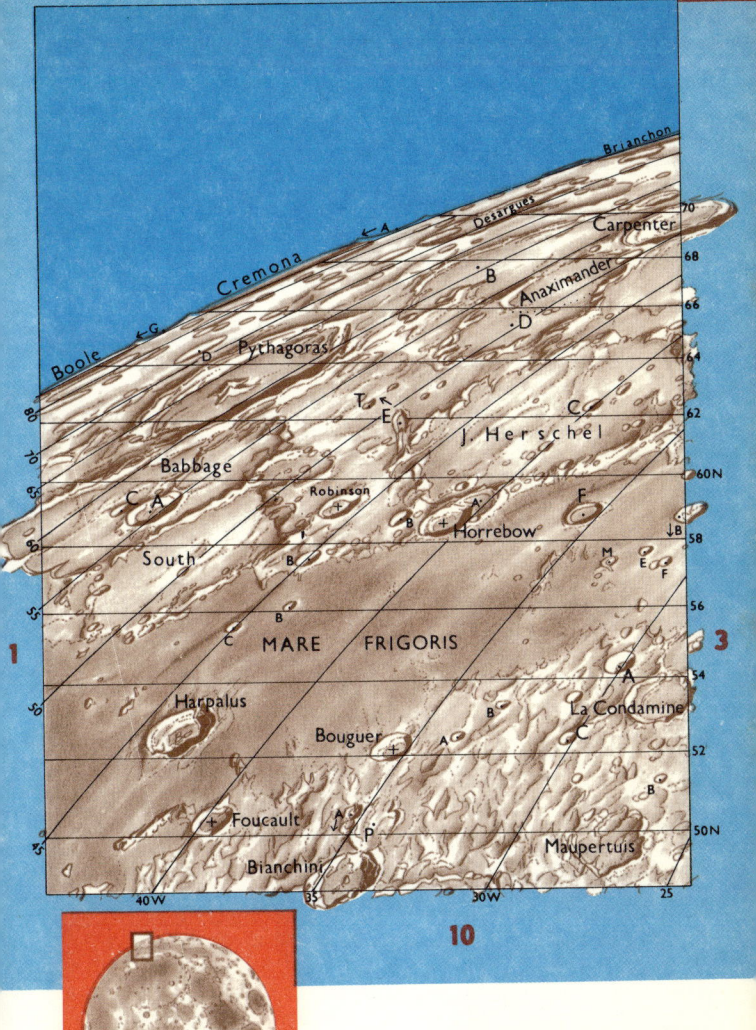

PYTHAGORAS

3 PLATO

Northern region of the Moon and the western part of Mare Frigoris. The lower part is occupied by the walled plain Plato with a very dark floor. Sea of Cold and Sea of Rains are separated by a narrow strip of continental area.

ANAXIMENES (44W, 72N) From Miletus, 585—528 BC; Greek philosopher; believed that the Earth was flat and that the Sun was hot because of the speed of its revolution round the Earth.
○ *crater (80 km)*

BRIANCHON (82W, 75N) Charles J. Brianchon, 1783—1864; French mathematician.
○ *crater in the libration zone (126 km)*

FONTENELLE (19W, 63N) Bernard le Bovier de Fontenelle, 1657—1757; French astronomer, popularizer of science, one of the early members of Académie des Sciences.
○ *crater (38 km)*

FRIGORIS, MARE Sea of Cold. Riccioli's name for an elongated mare in the northern polar region. The surface of M. Frigoris has an area of 440 000 sq km, comparable to that of the Black Sea on Earth.

MOUCHEZ (27W, 78N) Ernest A. B. Mouchez, 1821—92; French naval officer, later director of Paris observatory.
○ *remains of a crater (82 km)*

PASCAL (70W, 74N) Blaise Pascal, 1623—62; French mathematician, physicist, philosopher; invented an adding machine.
○ *crater (102 km)*

PHILOLAUS (32W, 72N) End of the 5th century BC; Greek philosopher, Pythagorean astronomer; developed theories of the Earth's motions and believed that the centre of the universe was a 'central fire'.
○ *crater (71 km)*

PLATO (9W, 51N) About 427—347 BC; prominent Greek philosopher, pupil of Socrates; adherent of Pythagorean astronomy, which maintains that the round Earth is surrounded by planetary spheres and other stars.
○ *walled plain (100 km)*

PONCELET (53W, 76N) Jean V. Poncelet, 1788—1867; French mathematician.
○ *crater (65 km)*

SYLVESTER (80W, 83N) James J. Sylvester, 1814—97; British mathematician; theory of numbers, analytical geometry.
○ *crater in the libration zone (58 km)*

Sylvester
80N
Mouchez
Brianchon
C
Poncelet
Pascal
75
D
G
Anaximenes
Philolaus
70
60
Carpenter
C
70N
E
B
50
68
E
A
45
66
40
C
F
64
D
Fontenelle
B
35
62
60N
MARE FRIGORIS
58
30
56
T
54
Z
B
Y
M
A
O
52
P
Plato
A
C
(Maupertuis)
50N
E
D
25
20W
15
10W

2

4

11

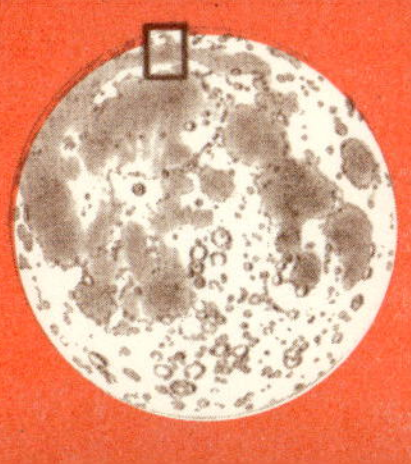

PLATO

4 ARCHYTAS

Area surrounding North Pole, central part of Mare Frigoris, northern part of the Alps.

ALPES, VALLIS Alpine Valley. Length 130 km; cleft in floor.
ANAXAGORAS (10W, 74N) 500−428 BC; Greek philosopher.
    ○ *crater with system of rays (51 km)*
ARCHYTAS (5E, 59N) About 428−347 BC; Greek philosopher, statesman, geometer.   ○ *crater (32 km)*
BARROW (8E, 71N) Isaac Barrow, 1630−77; English mathematician; Isaac Newton was one of his pupils.
    ○ *crater (93 km)*
BIRMINGHAM (11W, 65N) John Birmingham, 1829−84; Irish selenographer. ○ *remains of a crater (98 km)*
BOND, W. (4E, 65N) William C. Bond, 1789−1859; American astronomer.   ○ *walled plain (158 km)*
BYRD (10E, 85N) Richard E. Byrd, 1888−1957; American polar explorer and pilot.   ○ *walled plain (83 km)*
CHALLIS (9E, 80N) James Challis, 1803−62; English astronomer.   ○ *crater (56 km)*
EPIGENES (5W, 67N) 3rd century BC(?); Greek astronomer.
    ○ *crater (55 km)*
GIOJA (2E, 83N) Flavio Gioja, about 1302; Italian explorer.
    ○ *crater (42 km)*
GOLDSCHMIDT (3W, 73N) Hermann Goldschmidt, 1802−66; German amateur astronomer. ○ *walled plain (125 km)*
HERMITE (88W, 86N) Charles Hermite, 1822−1901; French mathematician.   ○ *crater (109 km)*
MAIN (10E, 81N) Robert Main, 1808−78; English astronomer.
    ○ *crater (51 km)*
METON (19E, 74N) About 432 BC; Greek astronomer.
    ○ *remains of a walled plain (122 km)*
PEARY (30E, 88N) Robert E. Peary, 1856−1920; American polar explorer.   ○ *walled plain (84 km)*
PROTAGORAS (7E, 56N) About 481−411 BC; Greek philosopher.   ○ *crater (22 km)*
SCORESBY (14E, 78N) William Scoresby, 1789−1857; English navigator, oceanographer.   ○ *crater (56 km)*
TIMAEUS (1W, 63N) About 400 BC; Pythagorean philosopher, contemporary with Plato. ○ *crater (33 km)*
TROUVELOT (6E, 49N) Etienne L. Trouvelot, 1827−95; French astronomer.   ○ *crater (9 km)*

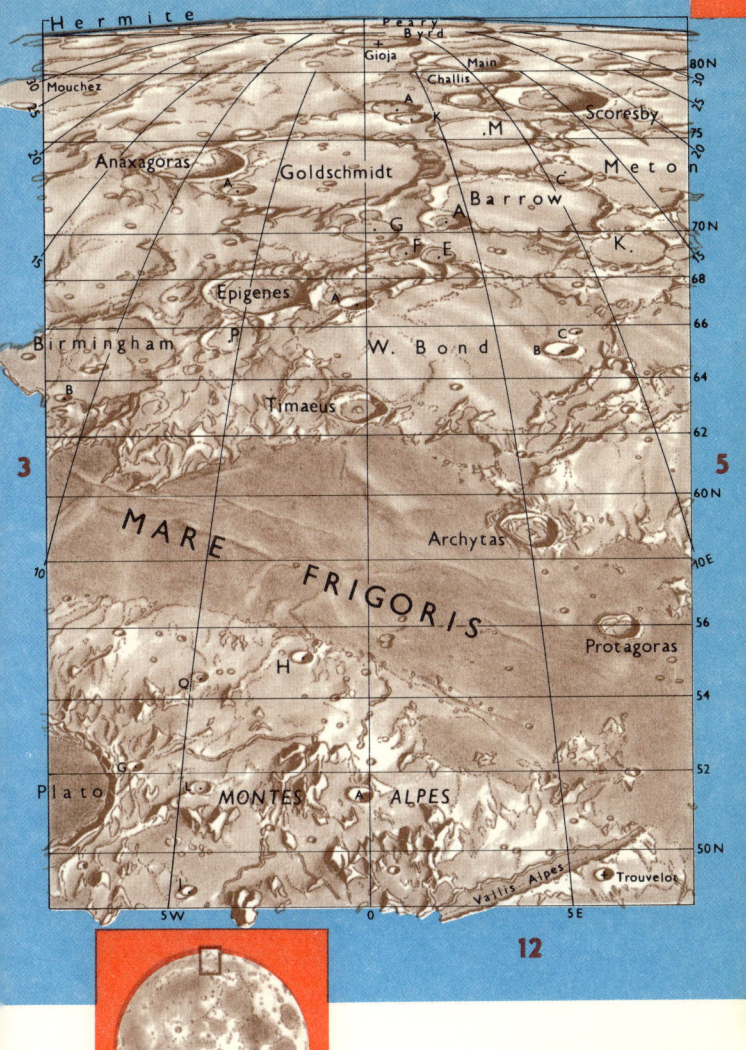

ARCHYTAS

5 ARISTOTELES

The eastern part of Mare Frigoris and the northern portion of the Moon. The most interesting formation of this area is the vast crater Aristoteles.

ARNOLD (36E, 67N) Christoph Arnold; German amateur astronomer. ○ *crater (95 km)*

ARISTOTELES (17E, 50N) 383–322 BC; Greek philosopher, polyhistor, whose teaching influenced European thinking for several centuries. ○ *crater, terrace-like walls (87 km)*

BAILLAUD (37E, 74N) Benjamin Baillaud, 1848–1934; French astronomer. ○ *flooded crater (89 km)*

DEMOCRITUS (35E, 62N) About 460–360 BC; Greek philosopher; put forward doctrine of atoms.
○ *prominent crater (39 km)*

DE SITTER (38E, 80N) Willem De Sitter, 1872–1934; notable Dutch astronomer. ○ *crater (65 km)*

EUCTEMON (31E, 76N) About 432 BC; Greek astronomer.
○ *crater (62 km)*

GALLE (22E, 56N) Johann H. Galle, 1812–1910; German astronomer; discovered Neptune according to Leverrier's calculation on 23 September 1846. ○ *crater (21 km)*

KANE (26E, 63N) Elisha K. Kane, 1820–57; American explorer.
○ *flooded crater (55 km)*

MAYER, C. (17E, 63N) Christian Mayer, 1719–83; German astronomer. ○ *prominent crater (38 km)*

MITCHELL (20E, 50N) Maria Mitchell, 1818–89; American astronomer. ○ *crater (30 km)*

MOIGNO (29E, 66N) Francois N. M. Moigno, 1804–84; French mathematician, physicist. ○ *crater (36 km)*

NANSEN (93E, 81N) Fridtjof Nansen, 1861–1930; Norwegian polar explorer. ○ *crater in the libration zone (110 km)*

NEISON (25E, 68N) Edmund Neison, 1849–1940; English selenographer. ○ *crater (53 km)*

PETERMANN (66E, 74N) August Petermann, 1822–78; German geographer. ○ *crater (73 km)*

PETERS (30E, 68N) Christian A. F. Peters, 1806–80; German astronomer. ○ *crater (15 km)*

SHEEPSHANKS (17E, 59N) Anne Sheepshanks, 1789–1876; a patron of and benefactor to astronomy. ○ *crater (24 km)*

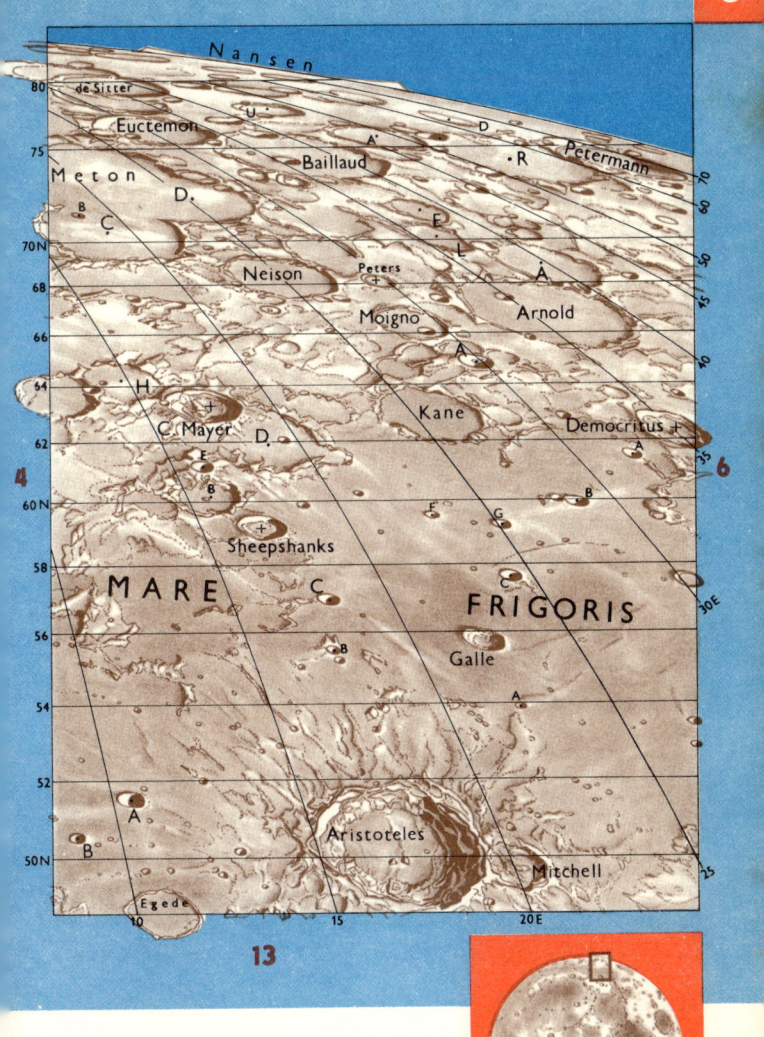

ARISTOTELES

6 STRABO

The map features a section of the north-eastern region of the Moon and the areas surrounding the eastern edge of Mare Frigoris. Important features for orientation are two craters, Strabo and Thales, and the prominent group of three craters, Strabo N, B and L.

BAILY (30E, 50N) Francis Baily, 1774—1844; English businessman, who from 1825 devoted himself fully to astronomy; observed the phenomenon now called 'Baily's beads' during an eclipse of the Sun in 1836.
○ *crater with disintegrated wall (27 km)*

CUSANUS (71E, 72N) Nikolaus Krebs Cusanus, 1401—64; German by origin, mathematician, cardinal; opposed the geocentric theory. ○ *crater with flooded floor (63 km)*

DE LA RUE (53E, 59N) Warren de la Rue, 1815—89; English astronomer, one of the pioneers in astrophotography.
○ *disintegrated walled plain (136 km)*

FRIGORIS, MARE Sea of Cold, see map 3.

GÄRTNER (35E, 59N) Christian Gärtner, 1750—1813; German mineralogist and geologist.
○ *remains of a walled plain (102 km)*

HAYN (83E, 65N) Friedrich Hayn, 1863—1928; German astronomer; improved upon the existing theory of the rotation of the Moon, mapped the limb areas of the Moon.
○ *crater (87 km)*

SCHWABE (46E, 65N) Heinrich Schwabe, 1789—1875; German astronomer; discovered the eleven-year cycle of solar activity.
○ *crater with flooded floor (25 km)*

STRABO (54E, 62N) About 54 BC — 24 AD; Greek geographer and historian; his 'Geografica' has remained one of the most significant works of its kind.
○ *prominent crater with terrace-like wall, flooded floor (55 km)*

THALES (50E, 62N) From Miletus, about 624—547 BC; founder of Greek geometry, philosopher; taught that water is the principle of everything.
○ *regular crater with sharp rim (32 km)*

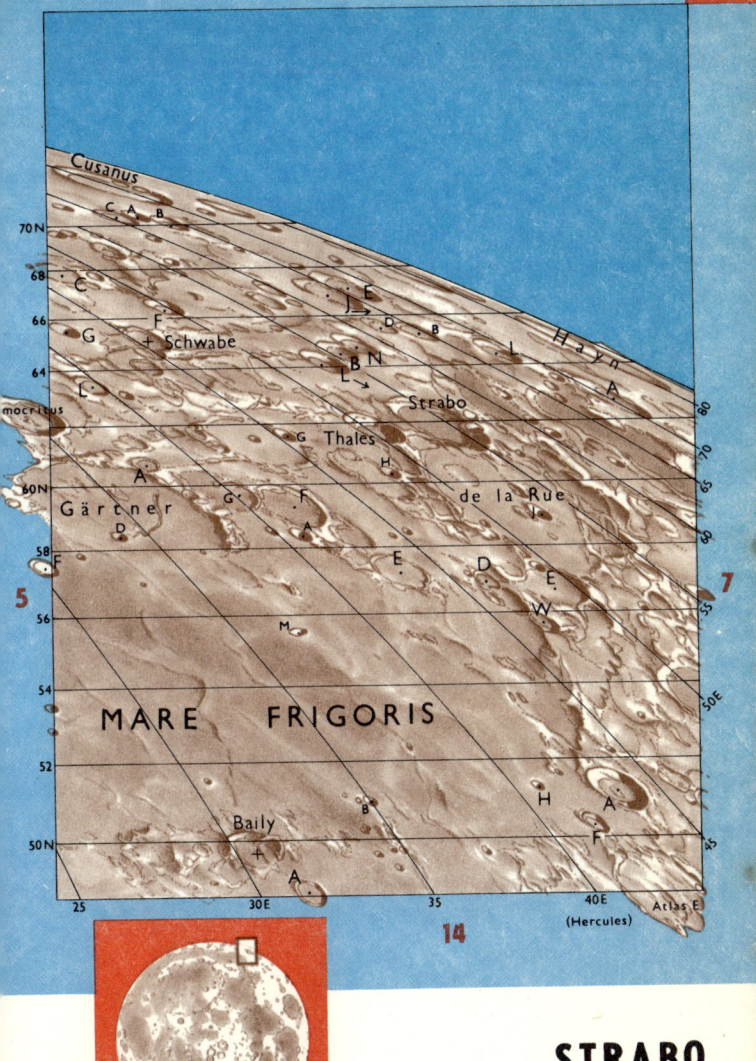

STRABO

7 ENDYMION

The north-eastern edge of the Moon has two very prominent dark areas: the floor of the flooded crater Endymion and the region of Mare Humboldtianum. In oblique illumination a mountain range stands out and forms an almost continuous wall around Humboldt's Sea. This wall is partly intersected by a wall of the crater Belkovich, which lies in the libration zone.

BELKOVICH (90E, 61N) Igor V. Belkovich, 1904—49; Soviet astronomer, specialist in selenodesy; observed and calculated the figure and rotational elements of the Moon.
○ *walled plain with central peaks and two larger craters on the periphery of the circular wall (198 km)*

ENDYMION (56E, 54N) A young shepherd who, according to a Greek legend, went to sleep on the mountain Latmos and whose beauty so aroused the cold heart of Diana or Selene (Goddess of the Moon) that she came down to the Earth to enjoy his company. ○ *very prominent crater, sizeable wall, flooded dark floor (125 km)*

HUMBOLDTIANUM, MARE Humboldt's Sea. Alexander von Humboldt, 1769—1859; German natural historian and explorer; in 1799 he observed the Leonid meteor shower; exploratory expeditions to the rivers Orinoco and Amazon in South America, and to Siberia. The name of Mare H. originates with Mädler, who based it symbolically on the fact that Mare H. joins the near and the far sides of the Moon, as Humboldt's discoveries linked the eastern and western hemispheres of the Earth.
○ *the eastern edge of Mare H. touches longitude 90 E and so the visibility of this Sea is considerably influenced by librations. Mare H. is a lunar basin, the outer limit of which has a diameter of 640 km. Its border stretches from the crater Strabo (see map 6) to the east of crater Endymion and out towards the south-east, around crater Mercurius E to the east, where it finally passes over to the far side.*

7

Belkovich
62
60N
MARE
B
58
C
HUMBOLDTIANUM
G F
56
G
6
54
J Endymion
E W D C A
52
D E B
L
G
M 50N
D P H E
45 (Atlas) 50E 55 60E 65 70E 75 80
15 (Mercurius)

ENDYMION

8 RÜMKER

This map is dominated by the dark surface of Oceanus Procellarum, the right edge of the map features an unusual elevated formation — Rümker. The Ocean of Storms is separated from the north-western limb of the Moon by a seemingly narrow strip of land.

ASTON (88W, 33N) Francis W. Aston, 1877—1945; British chemist and physicist, awarded Nobel Prize for chemistry in 1922; discovered 212 isotopes. ○ *crater (42 km)*
BUNSEN (86W, 42N) Robert W. Bunsen, 1811—99; German chemist; pioneer of chemical spectroscopy.
○ *disintegrated crater (61 km)*
DECHEN (68W, 46N) Ernst H. Karl von Dechen, 1800—89; German mineralogist and geologist. ○ *ring crater (12 km)*
GERARD (80W, 44N) Alexander Gerard, 1792—1839; Scottish explorer, known for his exploration of the Himalayas and Tibet.
○ *remains of a crater (74 km)*
HARDING (71W, 44N) Karl Ludwig Harding, 1765—1834; German astronomer; discovered asteroid Juno in 1804.
○ *crater with sharp walls (23 km)*
LAVOISIER (80W, 38N) Antoine Laurent Lavoisier, 1743—94; French chemist; one of the founders of modern chemistry.
○ *crater (68 km)*
LICHTENBERG (68W, 32N) Georg Ch. Lichtenberg, 1742—99; German physicist; worked in the field of static electricity; also known for his satirical reflections. ○ *crater (21 km)*
NAUMANN (62W, 35N) Karl Friedrich Naumann, 1797—1873; German geologist. ○ *small, sharp crater (9.6 km)*
PROCELLARUM, OCEANUS Ocean of Storms, see map 29.
RÜMKER (58W, 41N) Karl Ludwig Ch. Rümker, 1788—1862; German astronomer, director of naval school in Hamburg.
○ *solitary complex of lunar domes, the diameter of the formation is about 55 km*
ULUGH BEIGH (81W, 33N) 1393—1449; Uzbek astronomer and mathematician; founded astronomical school; built an observatory with a 40-m quandrant near Samarkand.
○ *disintegrated, flooded crater (56 km)*
Humason, Nielsen, see pp. 242, 243.

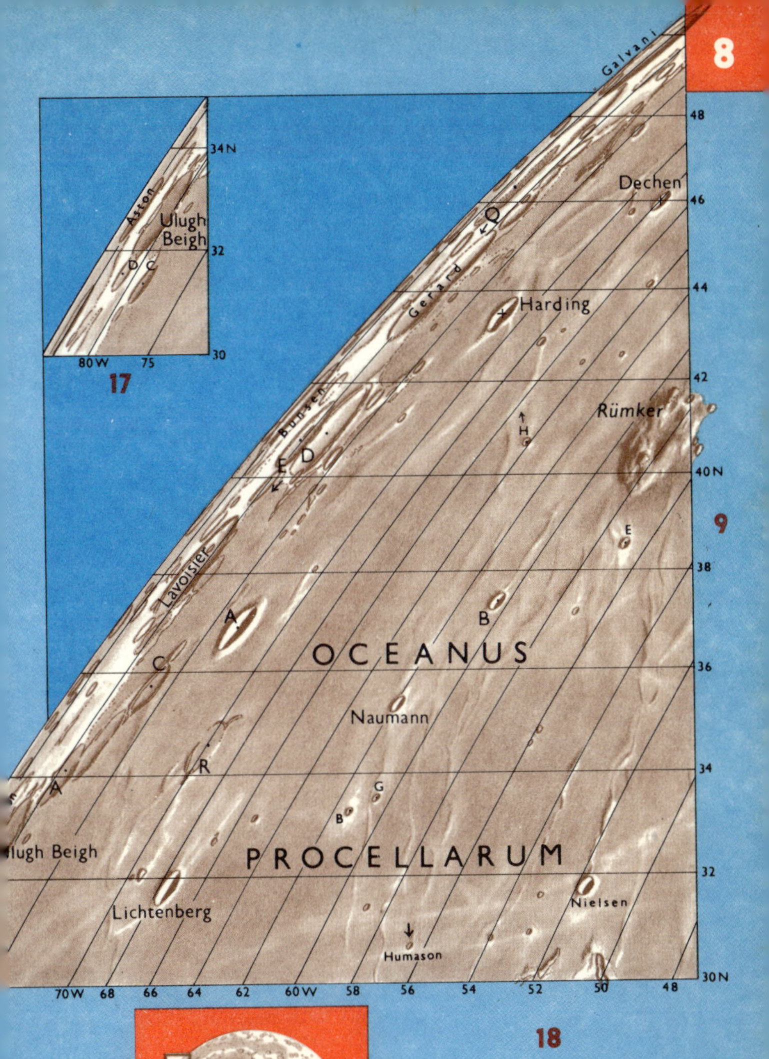
8
Galvani
48
34N
Aston
Ulugh
Beigh
32
D
C
30
80W
75
17
Dechen
46
Gerard
Harding
44
42
Rümker
Bunsen
H
E
D
40N
9
E
Lavoisier
38
A
B
OCEANUS
C
36
Naumann
R
34
A
G
B
lugh Beigh
PROCELLARUM
32
Lichtenberg
Nielsen
Humason
30N
70W
68
66
64
62
60W
58
56
54
52
50
48
18

RÜMKER

9 MAIRAN

Continent adjoining the Bay of Rainbows stretches here in to Oceanus Procellarum from north-east (c.f. map 10). A local curiosity is a group of peaks to the north of crater Gruithuisen. The formation Gruithuisen Gamma resembles an upturned bath tub; in reality it is a tall dome with a circular base about 19 km in diameter and with a crater (2 km) on its top.

DELISLE (35W, 30N) Joseph N. Delisle, 1688—1768; French astronomer; at the invitation of the Empress Catherine, he was put in charge of the new observatory in St Petersburg (1725—47); suggested a method for determining the distance of the Sun from observation of the transit of Mercury and Venus.
○ *crater (25 km/2 550 m)*

GRUITHUISEN (40W, 33N) Franz von Gruithuisen, 1774—1852; German physician and astronomer; dedicated but eccentric observer; wrote a book about his 'discoveries' of buildings and other evidence of life on the Moon.
○ *crater (15.2 km/1 860 m)*

IMBRIUM, MARE Sea of Rains, see map 11.

LOUVILLE (46W, 44N) Jacques d'Allonville, Chevalier de Louville, 1671—1732; French mathematician and astronomer; discovered a method for calculating exactly the path of the eclipse of the Sun.   ○ *eroded crater (36 km)*

MAIRAN (43W, 42N) Jean J. d'Ortons de Mairan, 1678—1771; French astronomer; studied aurora borealis.
○ *crater with sharp lip (41 km), Mairan T is a dome with peak crater*

PROCELLARUM, OCEANUS Ocean of Storms, see map 29.

RORIS, SINUS Bay of Dew. Riccioli's name for the northern promontory of Oceanus Procellarum. The Bay is partly intersected by a narrow cleft of Rima Sharp I, which can be seen in photographs taken by Lunar Orbiters between the craters Mairan T and D.

WOLLASTON (47W, 30N) William Hyde Wollaston, 1766—1828; English scientist of wide interests — medicine, chemistry, astronomy; discovered palladium and rhodium; devised a goniometer; in 1802 discovered dark lines in solar spectrum (subsequently rediscovered by Fraunhofer and which are now known as Fraunhofer's lines); Wollaston considered them to be merely dividing lines between colours in the spectrum.
○ *crater with sharp rim (10.2 km)*

MAIRAN

SINUS RORIS

OCEANUS PROCELLARUM

MARE IMBRUM

Sharp
Louville
Mairan
Wollaston
Ångström
Gruithuisen
Delisle

10 SINUS IRIDUM

The north-western part of Mare Imbrium with the beautiful Bay of Rainbows and the mountain range Jura, which forms its edge. Luna 17, which landed to the south of Cape Heraclides, transported an automatic mobile laboratory, Lunokhod I, to the Moon.

CARLINI (24W, 34N) Francesco Carlini, 1783—1862; Italian astronomer; worked in the field of celestial mechanics; improved the theory of the motion of the Moon.
○ *crater with sharp rim (11.4 km/2 200 m)*
HEIS (32W, 32N) Eduard Heis, 1806—77; German astronomer; observer of variable stars.
○ *crater (14 km/1 910 m), Heis A (6.1 km/650 m)*
HELICON (23W, 40N) 4th century BC; Greek mathematician and astronomer.
○ *crater (25 km/1 910 m), nearby is Helicon E (2.4 km/470 m)*
HERACLIDES, PROM. (34W, 41N) Cape Heraclides. Heraclides Ponticus, about 388—310 BC; pupil of Plato; taught that the Earth rotates on its axis.
HERSCHEL, C. (31W, 34N) Caroline Herschel, 1750—1848; worked for 50 years with her brother, William Herschel; discovered 8 comets. ○ *crater (13.4 km/1 850 m)*
IMBRIUM, MARE Sea of Rains, see map 11.
IRIDUM, SINUS Bay of Rainbows. Named by Riccioli.
○ *crater formation 260 km in diameter*
JURA, MONTES Jura Mountains. Named by Debes. The mountain range borders Sinus Iridum like the wall of a flooded crater.
LAPLACE, PROM. Cape Laplace. Pierre Simon Laplace, 1749—1827; eminent French mathematician, disciple of Newton; worked in the field of celestial mechanics (his great work 'Mécanique Céleste'); theory of the origin of the solar system.
SHARP (40W, 46N) Abraham Sharp, 1651—1742; English astronomer, assistant to Flamsteed at Greenwich Observatory.
○ *crater (40 km)*

SINUS IRIDUM

11 LE VERRIER

The central and northern part of Mare Imbrium is very poor in large craters and wrinkle ridges occur only in small numbers. During sunrise and sunset interesting views of the mountains along the northern edge of the map can be observed. They cast long, pointed shadows, which in the past conveyed a false impression of the height of these lunar mountains.

IMBRIUM, MARE Sea of Rains. Named by Riccioli. Its surface area is 860 000 sq km and after Oceanus Procellarum it is the second largest mare on the Moon, and is the largest lunar basin. M. Imbrium is surrounded by a circular mountain range which opens in the west as it joins Oceanus Procellarum. The perimeter of M. Imbrium is formed by the mountain ranges of Jura, Alps, Caucasus, Apennines and Carpathians. Tectonic clefts, which originated during the formation of the basin, can be traced far to the south and south-east. The diameter of the basin is 1 250 km. Remains of the inner wall of the basin are formed by Mts Recti, Mts Teneriffe, Pico and Mts Spitzbergensis. One of the first mascons was discovered in the centre of this basin.

LE VERRIER (21W, 40N) Urbain Jean Joseph Le Verrier, 1811—77; French mathematician and astronomer; independently of J. C. Adams calculated the orbit and position of Neptune. ○ *crater (21 km/2 100 m)*

PICO Mountain named by Schröter, who evidently had in mind 'Pico von Teneriffa'; he compared the height of this mountain with that of other lunar mountain ranges. Mount Pico is 2 400 m high; its base measures 15 × 20 km.

RECTI, MONTES Straight Range. Named thus by Birt because of its shape. Its length is almost 80 km, and its height is 1 800 m.

TENERIFFE, MONTES Teneriffe Mountains. Its name is reminiscent of the mountain Teneriffe, where Piazzi Smyth first tested telescopic observational conditions at high levels above the sea. On the Moon the peaks of Teneriffe reach a height of 2 400 m.

Small craters: Le Verrier D (9.1 km/1 830 m)
Le Verrier B (5.1 km/980 m)
Le Verrier W (3.3 km/620 m)

LE VERRIER

MONTES RECTI
MONTES TENERIFFE
Pico
MARE IMBRIUM
Le Verrier
(Carlini)
(Timocharis)
(MONTES SPITZBERGENSIS)

12 ARISTILLUS

The eastern edge of Mare Imbrium is one of the most interesting parts of the lunar surface. The lunar Alps with their well-known valley, the solitary mountain Piton and the group of three craters Archimedes, Autolycus and Aristillus are suitable objects for telescopic observation.

ALPES, MONTES Alps. This mountain range was named by Hevelius. Its height is 1 800 − 2 400 m.

AGASSIZ, PROM. Cape Agassiz. Louis J. R. Agassiz, 1807−73; Swiss naturalist.

BLANC, MONS Mont Blanc. A mountain 3 600 m high.

DEVILLE, PROM. Cape Deville. Sauncte-Claire Ch. Deville, 1814−76; French geologist.

ARISTILLUS (1E, 34N) About 280 BC; one of the earliest astronomers of the Greek Alexandrean school.
○ *crater with radiating rays (55 km/3 650 m) with a group of three peaks on the floor (900 m)*

AUTOLYCUS (1E, 31N) About 330 BC; Greek astronomer.
○ *crater (39 km/3 430 m)*

CASSINI (4E, 40N) 1. Giovanni Domenico Cassini, 1625−1712; Italian astronomer settled in France; discovered four of Saturn's satellites and the division in Saturn's ring now known as the 'Cassini Division'. 2. Jacques J. Cassini, 1677−1756; son of Domenico C., whom he succeeded as Director of the Observatory at Paris.
○ *flooded crater (56 km/1 240 m); Cassini A (17 km/2 830 m)*

KIRCH (6 W, 39 N) Gottfried Kirch, 1639−1710; German astronomer, discovered a large comet in 1680.
○ *sharp crater (11.7 km/1 830 m)*

LUNICUS, SINUS (1.5W, 32N) Location of the first touchdown of a space probe (Luna 2, 1959). The name was introduced by the IAU in 1970.

PIAZZI SMYTH (3W, 42N) Charles Piazzi Smyth, 1819−1900; Astronomer Royal for Scotland.
○ *sharp crater (12.8 km/2 530 m)*

PITON Mount Piton. Named after a peak in the Teneriffe massif.
○ *an isolated mountain 2 250 m high*

SPITZBERGENSIS, MONTES Spitsbergen. Mountain range, so named by M. Blagg because of its similar shape to that of terrestrial Spitsbergen. ○ *mountains, 1 500 m in height*

THEAETETUS (6E, 37N) About 380 BC; Athenian philosopher, friend of Plato (it is near Plato on the Moon).
○ *crater (25 km/2 830 m)*

4
11
13
22
6W
4
2
0
2
4
6E
48
46
44
42
40N
38
36
34
32
30N
Trouvelot
Alpes
Vallis
Pico C
K
MONTES ALPES
Mons Blanc
M
G
L
Pr. Deville
Pr. Agassiz
E
W
Y
Piazzi Smyth
Piton
U
B
A
M
Cassini
A
B
B
Kirch
K
F
γ
MARE IMBRIUM
Theatetus
MONTES SPITZBERGENSIS
B
Aristillus
A
SINUS LUNICUS
D
C
Autolycus
A
K
T
Luna 2
S
Archimedes

ARISTILLUS

13 EUDOXUS

The left part of the map is occupied by the Caucasus Mountains range, which lie on the boundary of the Sea of Rains and the Sea of Serenity. The north-western part of the Sea of Serenity, with its bright radiating rays from crater Aristillus (map 12), is situated at the base of the map. The large crater Eudoxus forms a striking companion with Aristoteles (map 5).

ALEXANDER (13E, 40N) Alexander the Great of Macedonia, 356 – 323 BC; statesman, commander; his expeditions broadened Greek knowledge of the Earth; the city of Alexandria in Egypt became a centre of science.
○ *greatly eroded walled plain (82 km)*

CALIPPUS (11E, 39N) About 330 BC; Greek astronomer, pupil of Eudoxus; improved the system of homocentric spheres.
○ *crater (31 km/2 690 m)*

CAUCASUS, MONTES Caucasus Mountains. Named by Mädler. This is a direct continuation of the lunar Apennines, from which the Caucasus is separated by a mere 50 km-wide channel between the Sea of Rains and the Sea of Serenity.

EGEDE (11E, 49N) Hans Egede, 1686 – 1758; Danish missionary, who worked for 15 years in Greenland.
○ *flooded crater with a low wall (37 km)*

EUDOXUS (16E, 44N) About 408 – 355 BC; famous Greek astronomer, a student of Plato; eminent geometer; devised a system of 7 spheres concentric to the Earth and rotating uniformly about different axes to account mathematically for the motions of celestial bodies.
○ *prominent crater with a sharp rim (67 km)*

LAMÈCH (13E, 43N) Felix Ch. Lamèch, 1894 – 1962; French astronomer, selenographer.
○ *sharp crater (13 km/1 460 m)*

SERENITATIS, MARE Sea of Serenity, see map 24.
  Small craters: Linné H (3.2 km/730 m)
              Linné F (5.0 km/1 050 m)
              Eudoxus D (9.6 km/1 300 m)
              Cassini C (13.7 km/2 420 m)

(Aristoteles)

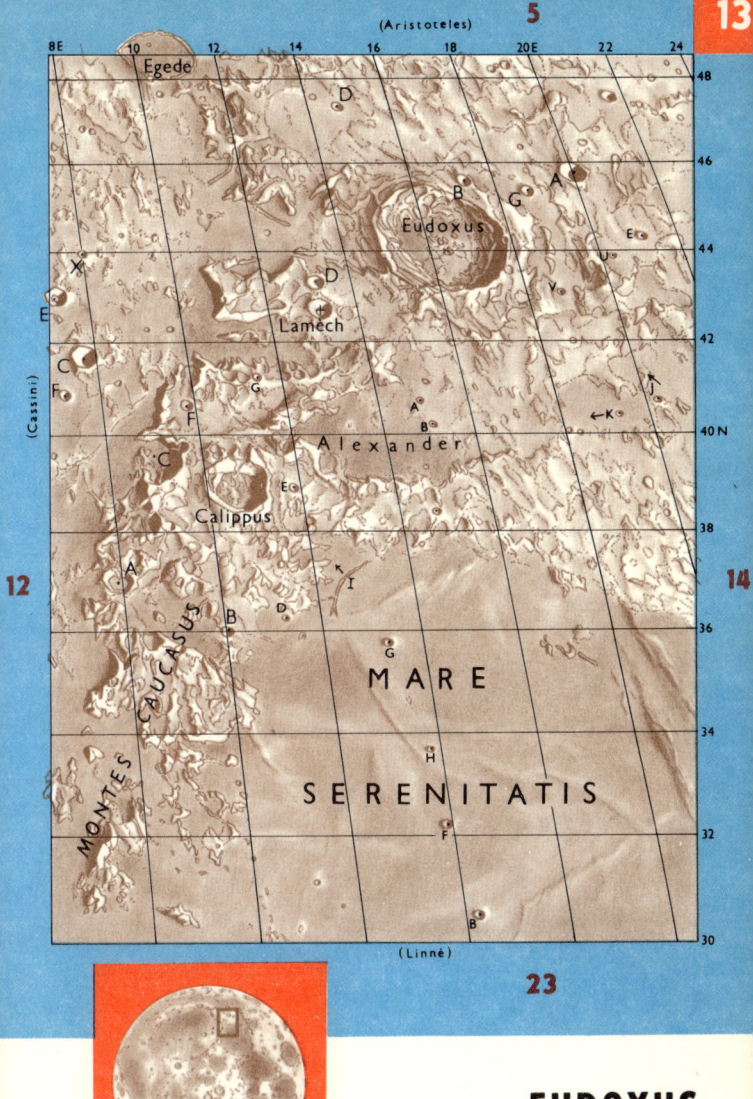

Egede

Eudoxus

Lamech

Alexander

Calippus

MONTES CAUCASUS

MARE

SERENITATIS

(Cassini)

(Linné)

EUDOXUS

14 HERCULES

A very interesting landscape with a great variety of lunar features. Clefts in crater Bürg, in the floor of Posidonius and Rimae Daniell. Dark areas of Lacus Mortis, Lacus Somniorum and north-eastern edge of Mare Serenitatis.

BÜRG (28E, 45N) Johann Tobias Bürg, 1766—1834; Austrian astronomer; theory of the motion of the Moon.
○ *prominent crater in Lacus Mortis (40 km)*
CHACORNAC (32E, 30N) Jean Chacornac, 1823—73; French astronomer; discovered 6 small planets.
○ *crater with a disintegrated wall (51 km/1 450 m)*
DANIELL (31E, 35N) John Frederick Daniell, 1790—1845; English chemist and meteorologist; invented hygrometer.
○ *oval crater (30 × 23 km/2 070 m)*
GROVE (33E, 40N) Sir William Robert Grove, 1811—96; English lawyer, who in his spare time carried out valuable research in physics. ○ *crater (27 km)*
HERCULES Hero of Greek mythology gifted with superhuman strength. Riccioli presumed that he was an astronomer who lived about 1560 BC.
○ *prominent crater with dark areas on its floor (67 km)*
LUTHER (24E, 33N) Robert Luther, 1822—1900; German astronomer, discovered 24 small planets.
○ *crater (9.5 km/1 900 m)*
MASON (30E, 43N) Charles Mason, 1730—87; English astronomer; assistant in Greenwich Observatory; patient observer.
○ *flooded, partly disintegrated crater (33 × 43 km)*
MORTIS, LACUS Lake of Death. Named by Riccioli. A formation 162 km in diameter; resembles a flooded crater; clefts and furrows in its surface.
PLANA (28E, 42N) Giovanni A. A. Plana, 1781—1864; Italian astronomer and mathematician.
○ *crater with a central peak (44 km)*
POSIDONIUS (30E, 32N) 135—51 BC; Greek philosopher, geographer, astronomer.
○ *prominent walled plain (100 km/2 300 m)*
SERENITATIS, MARE Sea of Serenity, see map 24.
SOMNIORUM, LACUS Lake of Dreams. Named by Riccioli. Irregular contours and indefinite borders.
WILLIAMS (37E, 42N) Arthur Stanley Williams, 1861—1938; English lawyer, a diligent observer of the planets, especially Jupiter. ○ *disintegrated crater (36 km)*

HERCULES

15 ATLAS

Area close to the north-eastern margin of the Moon. Eastern promontory of Lacus Somniorum with a wide cleft G. Bond I. Hills along the base of the map link up with the mountain chain Taurus.

ATLAS (44E, 47N) According to Greek mythology, one of the Titans. According to Riccioli, Atlas was a Moroccan king, who lived about 1580 BC. ○ *crater (87 km)*

BERZELIUS (51E, 36N) Jöns Jacob Berzelius, 1779—1848; Swedish chemist; author of modern chemical notation.
○ *crater (51 km)*

BOND, G. (36E, 32N) George P. Bond, 1826—65; American astronomer; demonstrated that the rings of Saturn cannot be solid. ○ *crater (20 km/2 780 m)*

CARRINGTON (62E, 44N) Richard Ch. Carrington, 1826—75; English astronomer; determined the rotation of the Sun; observed the solar flare on the great sunspot of 1859.
○ *crater (30 km)*

CEPHEUS (46E, 41 N) Mythological king of Ethiopia, whose name was given to a constellation. ○ *crater (40 km)*

CHEVALLIER (51E, 45N) Temple Chevallier, 1794—1873; French by origin; mathematician; Director of Durnham Observatory in England. ○ *disintegrated, flooded crater (52 km)*

FRANKLIN (48E, 39N) Benjamin Franklin, 1706—90; American statesman and scientist; invented lightning-conductor.
○ *prominent crater (56 km)*

HALL (37E, 34N) Asaph Hall, 1829—1907; American astronomer; discovered the satellites of Mars.
○ *disintegrated, flooded crater (39 km/1 140 m)*

HOOKE (55E, 41N) Robert Hooke, 1635—1703; English natural philosopher, inventor. ○ *flooded crater (37 km)*

KIRCHHOFF (39E, 30N) Gustav Robert Kirchhoff, 1824—87; German physicist; developed the basic principles of spectrum analysis. ○ *crater (25 km/2 590 m)*

MAURY (40E, 37N) 1. Matthew F. Maury, 1806—73; American oceanographer. 2. Antonia C. Maury, 1866—1952; American astronomer. ○ *crater (17.6 km/3 270 m)*

MERCURIUS (66E, 46N) Mercury, legendary messenger of the Gods. ○ *crater (68 km)*

OERSTED (47E, 43N) Hans Christian Oersted, 1777—1851; Danish physicist and philosopher. ○ *flooded crater (42 km)*

SHUCKBURGH (53E, 43N) Sir George Shuckburgh, 1751—1804; English astronomer. ○ *crater (38 km)*

ATLAS

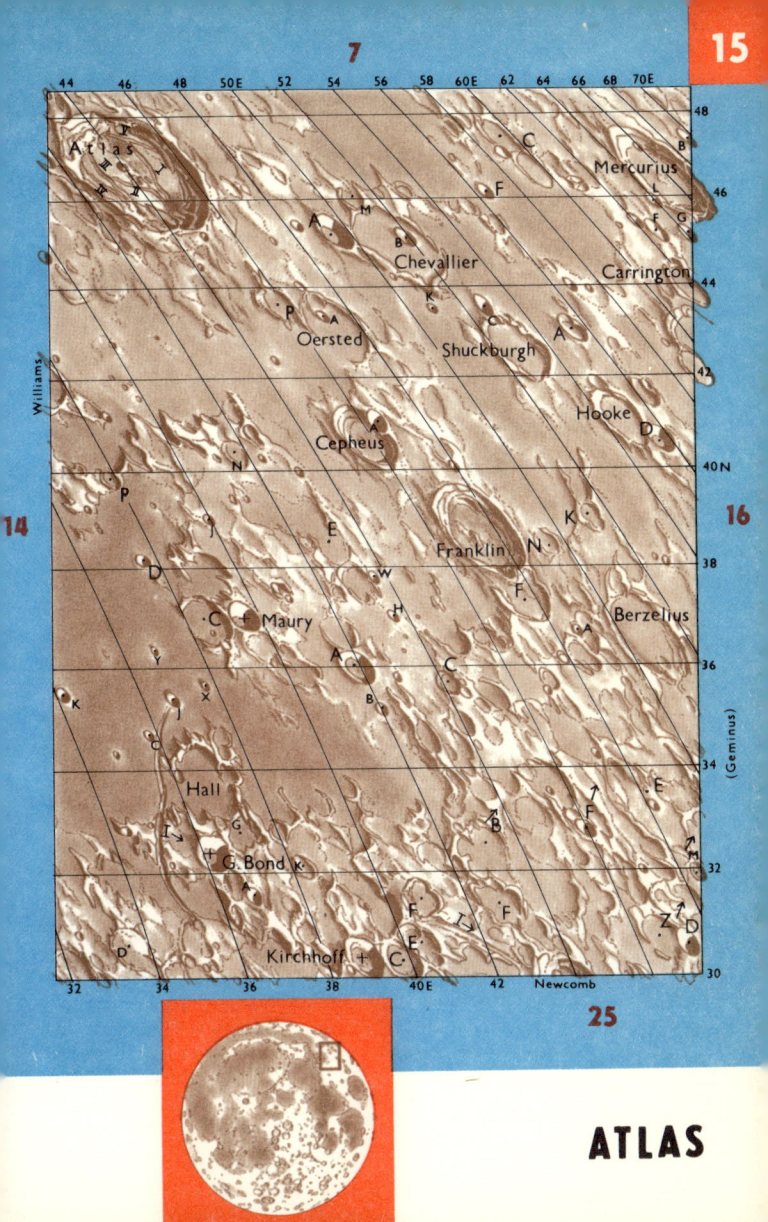

16 GAUSS

The north-eastern limb of the Moon containing the large walled plain Gauss and the prominent craters Geminus, Berosus and Hahn. To the east of crater Schumacher is a dark patch resembling a mare area.

BERNOUILLI (61E, 35N) 1. Jacques Bernouilli, 1654—1705 2. Jean Bernouilli, 1667—1784. Two brothers; Swiss mathematicians. ○ *crater (47 km)*

BEROSUS (70E, 34N) Berosus of Chaldea, about 270 BC; Babylonian priest, historian, astronomer; noted the locked rotation of the Moon. ○ *flooded crater (74 km)*

BOSS (88E, 46N) Lewis Boss, 1846—1912; American astronomer, author of positional catalogues of the stars.
○ *crater (47 km)*

BURCKHARDT (56E, 31N) Johann Karl Burckhardt, 1773—1825; German astronomer, worked in time-measuring service.
○ *crater (57 km)*

GAUSS (79E, 36N) Karl Friedrich Gauss, 1777—1855; famous German mathematician and physicist, geodesist and theoretical astronomer; Director of Observatory in Göttingen; invented magnetometer. ○ *walled plain (177 km)*

GEMINUS (57E, 34N) About 70 BC; Greek astronomer.
○ *prominent crater (86 km)*

HAHN (74E, 31N) Friedrich, Graf von Hahn, 1741—1805; German amateur astronomer; diligent observer.
○ *crater (84 km)*

MESSALA (60E, 39N) Ma-sa-Allah (or Mashalla), died about 815 AD; Jewish astronomer and astrologer; author of textbooks which were still used in Europe during the Middle Ages.
○ *walled plain (124 km)*

RIEMANN (40N, 88E) Georg Friedrich Bernhard Riemann, 1826—66; German mathematician; developed 'Riemann's geometry' and geometrical foundations of modern physics.
○ *remains of walled plain (133 km)*

SCHUMACHER (61E, 42N) Heinrich Christian Schumacher, 1780—1850; German astronomer of Danish origin; founder of the specialist periodical 'Astronomische Nachrichten'.
○ *eroded, flooded crater (61 km)*

ZENO (73E, 45N) About 340—264 BC; Cypriot astronomer, who correctly explained the reasons for the eclipse of the Sun and the Moon. ○ *crater (65 km)*

GAUSS

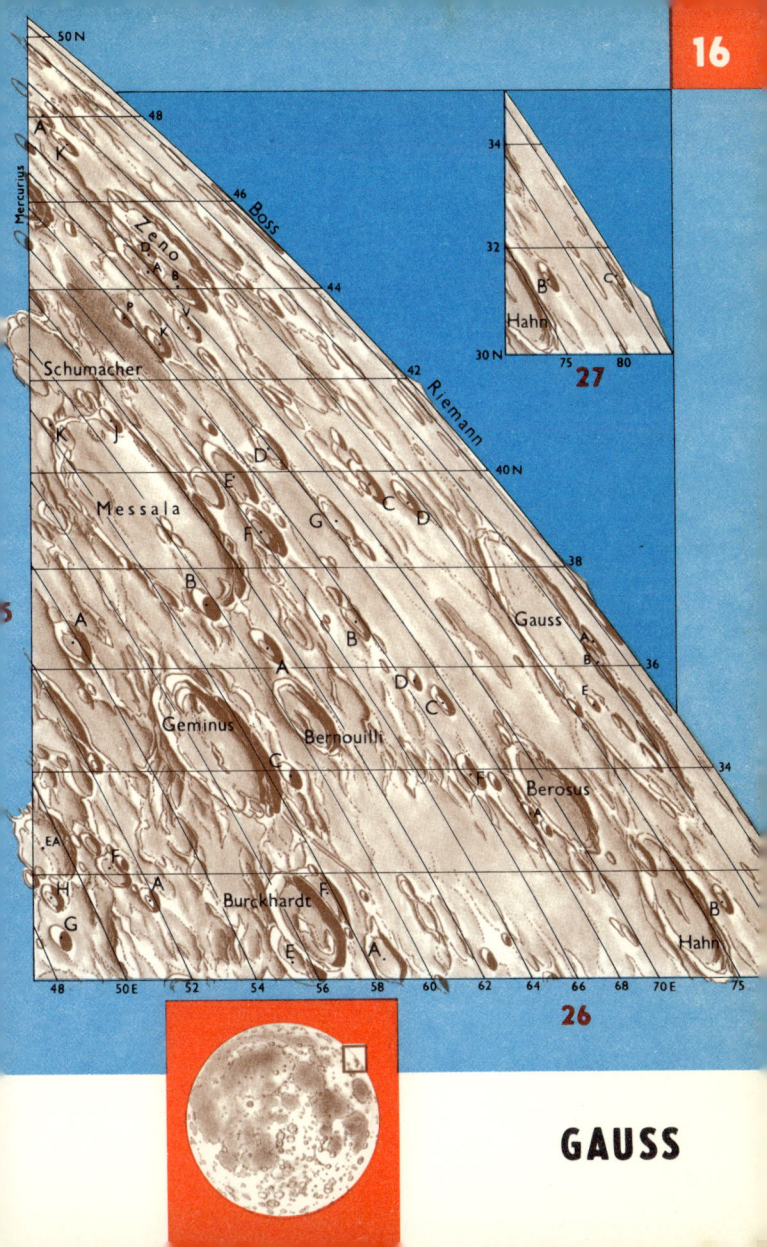

17 STRUVE

The western edge of Oceanus Procellarum and the western margin of the Moon. Bright rays radiate from crater Olbers, the landing place of the Soviet automatic probe Luna 13.

BALBOA (83W, 19N) Vasco Nuñez de Balboa, about 1475 – 1517; Spanish traveller and conquistador; the first European to sight and reach the Pacific Ocean. ○ *flooded crater (73 km)*

BARLELS (90W, 24N) Julius Barlels, 1899 – 1964; German geophysicist. ○ *crater (45 km)*

BRIGGS (69W, 26N) Henry Briggs, 1556 – 1630; English mathematician; devised a system of logarithms.
○ *crater (39 km)*

DALTON (17N, 84W) John Dalton, 1766 – 1844; English chemist and physicist. ○ *crater (58 km)*

EDDINGTON (72W, 22N) Sir Arthur Stanley Eddington, 1822 – 1944; prominent English astrophysicist and mathematician; determined the internal structure of stars, studied relativity. ○ *remains of flooded walled plain (134 km)*

EINSTEIN (88W, 17N) Partly extends behind 90W. Albert Einstein, 1879 – 1955; renowned physicist; author of the general and special theories of relativity.
○ *walled plain (190 km) with a central crater (45 km)*

KRAFFT (73W, 17N) Wolfgang Ludwig Krafft, 1743 – 1814; astronomer and physicist of German origin; worked all his life in St Petersburg. ○ *crater with a flooded floor (51 km)*

RUSSELL 1. John Russell, 1745 – 1806; English painter and amateur astronomer. 2. Henry Norris Russell, 1877 – 1957; American astronomer, co-author of Hertzsprung-Russell diagram. ○ *remains of a walled plain (99 km)*

SELEUCUS (66W, 21N) About 150 BC; Babylonian astronomer.
○ *prominent crater (43 km)*

STRUVE (77W, 24N) 1. Friedrich Georg Wilhelm von Struve, 1783 – 1864; German-Russian astronomer; Director of Pulkovo Observatory; observed double stars and the parallax of stars. 2. Otto Wilhelm von Struve, 1819 – 1905; son of the former, Director of Pulkovo Observatory; discovered a satellite of Uranus. 3. Otto Struve, 1897 – 1963; Russo – American astrophysicist, grandson of the former.
○ *remains of a flooded walled plain (183 km)*

VOSKRESENSKY (88W, 28N) Leonid A. Voskresensky, 1913 – 65; eminent Soviet specialist in the field of rockets.
○ *flooded crater (47 km)*

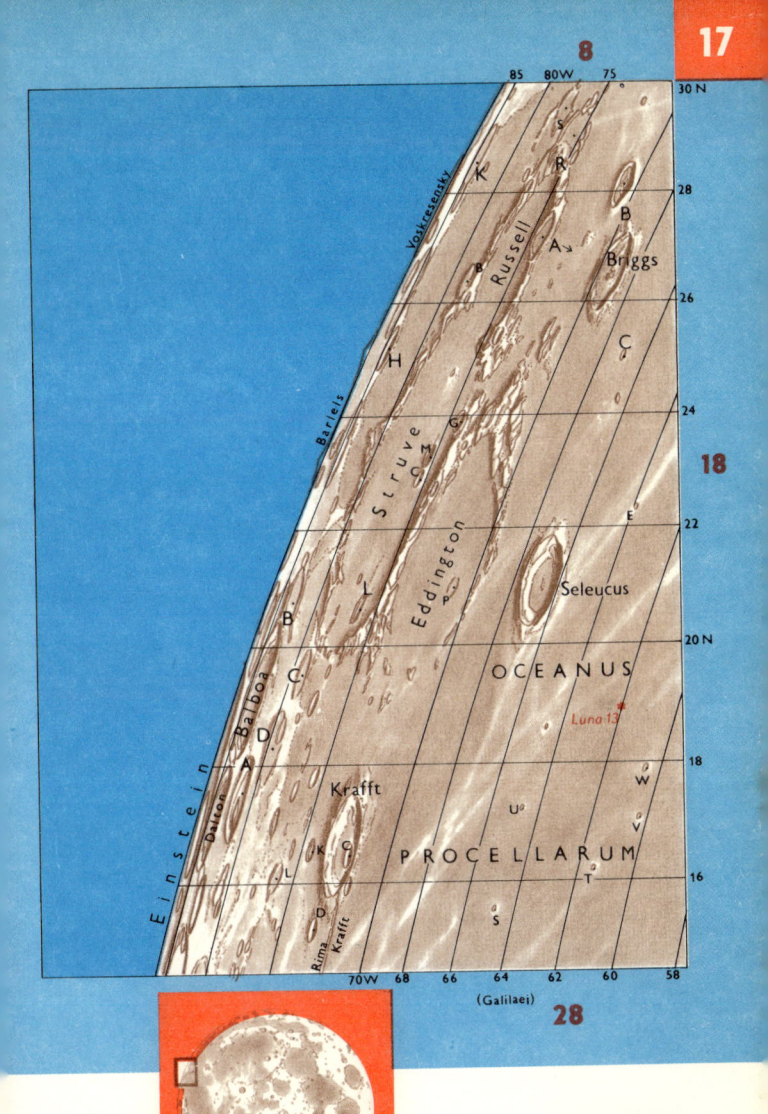

STRUVE

18 ARISTARCHUS

Exceptionally interesting features can be observed in this part of Oceanus Procellarum. These include Schröter's valley, which is visible even through a small telescope, and the long, sinuous rilles Rima Marius and Rimae Aristarchus. Herodotus Omega is a lunar dome in the area of crater Marius (c.f. map 29). Crater Aristarchus is the most distinctive in this area.

ARISTARCHUS (48W, 24N) About 310—230 BC; Greek astronomer from Samos; first to teach that the Earth revolves round the Sun and rotates on its axis.
○ *exceptionally bright crater, visible even in earthshine, and on the dark portion of the crescent Moon; centre of bright rays (45 km/3630 m)*

HERODOTUS (50W, 23N) About 484—408 BC; Greek historian; 'father of history'. ○ *flooded crater (35 km)*

MARIUS, RIMA A typical sinuous rille, named after the nearby crater Marius (see map 29). The rille starts about 25 km northwest of crater Marius C, at which point it is about 2 km wide. It wends north and by crater Marius B it turns west and narrows to 1 km. The rille ends about 40 km west of crater Marius P, where its width is only about 500 m. Total length of the rille is about 250 km.

SCHIAPARELLI (59W, 23N) Giovanni V. Schiaparelli, 1835—1910; Italian astronomer, who developed terminology used in the maps of Mars; in 1877 discovered 'canals' on Mars. Founder of modern meteoric astronomy. ○ *crater (24 km)*

VALLIS SCHRÖTERI The largest sinuous valley-cleft on the Moon, named after German selenographer Schröter (crater, see map 32). The valley starts 25 km north of crater Herodotus. It resembles a dry river bed with numerous meanders and starts from a crater measuring 6 km in diameter and widens to 10 km. Its beginning is shaped like a 'Cobra Head'. It gradually narrows to 500 m and terminates on the edge of a tetragonal land mass. The total length of the valley is 200 km. The floor of the valley is flat and merges into another sinuous rille, which is not visible from the Earth. The depth of the valley is about 1 000 m, but it diminishes towards its end.

Freud, Väisälä, Zinner, see pp. 242, 243.

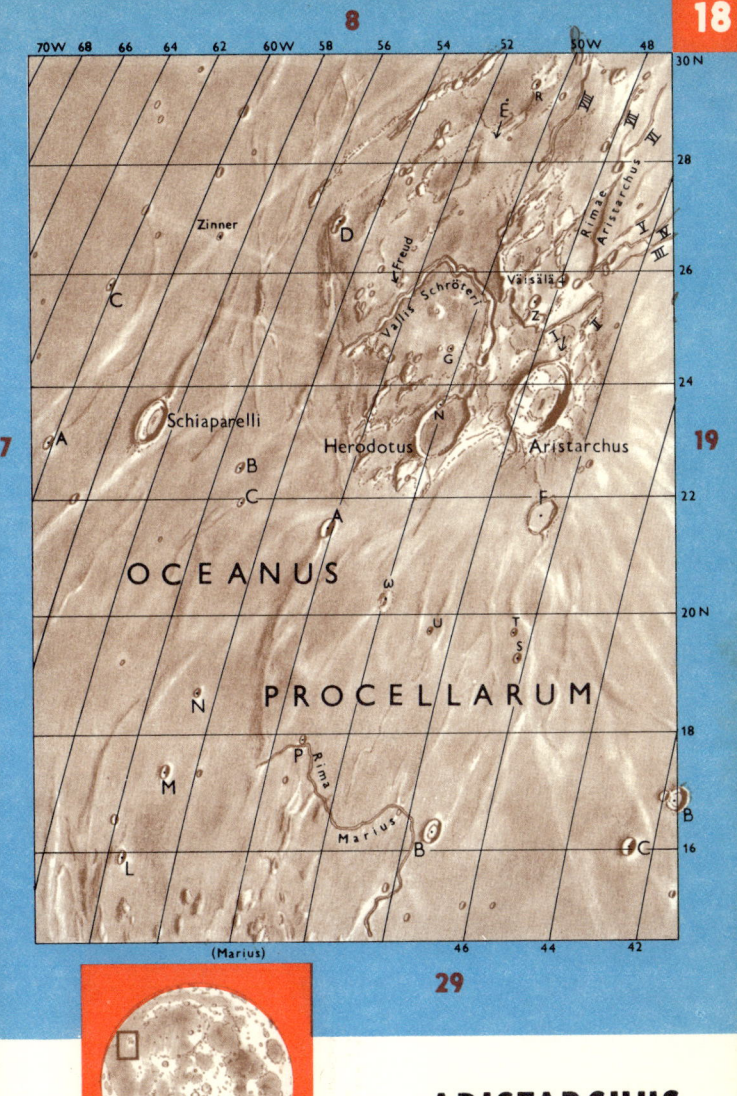

ARISTARCHUS

19 BRAYLEY

The south-western part of Mare Imbrium with numerous wrinkle ridges and a network of bright rays radiating from craters Copernicus and Aristarchus. System of sinuous rilles in the vicinity of crater Prinz provides a very interesting object for observation with large telescopes.

ÅNGSTRÖM (42W, 30N) Anders J. Ångström, 1814—74; Swedish physicist; the Ångström is a unit of wavelength named in his honour. ○ *crater (9.8 km/2 030 m)*
BESSARION (37W, 15N) Johannes Bessarion, died 1472; Cardinal of Greek origin, lived in Italy; invited Regiomontanus to Italy. ○ *crater (10.2 km/2 000 m)*
BRAYLEY (37W, 21N) Edward William Brayley, 1802—70; English writer on science; observed bolides.
○ *crater (14.5 km/2 840 m)*
DELISLE, see map 9.
DIOPHANTUS (34W, 28N) About 4th century AD; prominent mathematician of Alexandria; established principles for solving equations. ○ *crater (18.5 km/2 970 m)*
HARBINGER, MONTES Harbinger Mountains. Harbingers of dawn on Aristarchus. A mountain chain which appears with the rising Sun and heralds the dawn above crater Aristarchus. This is a group of isolated peaks on the edge of Mare Imbrium.
IMBRIUM, MARE Sea of Showers, see map 11.
KRIEGER (46W, 29N) Johann N. Krieger, 1865—1902; German selenographer; entered the details of the lunar surface on enlarged photographs.
○ *flooded crater (22 km). Wall is broken up by crater Krieger B (9.5 km)*
MAYER, TOBIAS (29W, 16N) Tobias Mayer, 1723—62; German selenographer; produced an accurate map of the Moon.
○ *crater (33 km/2 920 m)*
PRINZ (44W, 26N) Wilhelm Prinz, 1857—1910; German selenologist; undertook comparative studies of the lunar and terrestrial surfaces.
○ *remains of a flooded crater (52 km/1 010 m); one of the sinuous rilles resembling a dry river bed originates in crater Prinz A (4.9 km/180 m)*
PROCELLARUM, OCEANUS Ocean of Storms, see map 29.

BRAYLEY

20 PYTHEAS

The southern part of Mare Imbrium, bordered by the mountain chain of Montes Carpatus (see also map 31). Complex structure of bright rays in this area belongs to the circle of rays surrounding crater Copernicus (see map 31). Interesting objects for observation are the crater chain originating at bottom right of the map and the 'ghost' crater Lambert R, visible only when close to the terminator.

CARPATUS, MONTES Carpathian Mountains. Mädler's name for a characteristic lunar mountain range which follows the southern border of Mare Imbrium.

DRAPER (22W, 18N) Henry Draper, 1837–82; American astronomer; one of the pioneers of astrophotography and spectroscopy; photographed the Moon and spectra of the stars; first to photograph the nebula in Orion.
○ *crater (8.8 km/1 740 m)*

EULER (29W, 23N) Leonard Euler, 1707–83; Swiss mathematician; worked in pure and applied mathematics and celestial mechanics. ○ *crater (28 km/2 240 m)*

IMBRIUM, MARE Sea of Rains, see map 11.

LA HIRE (MONS) (26W, 28N) Philippe de La Hire, 1640–1718; French mathematician, engineer, astronomer.
○ *isolated mountainous massif measuring about 10×20 km*

LAMBERT (21W, 26N) Johann Heinrich Lambert, 1728–77; German mathematician and astronomer.
○ *prominent crater with terrace-like walls (30 km/2 690 m)*

PYTHEAS (21W, 20N) From Athens, about 350 BC; Greek navigator who sailed far to the north, where the Sun rose 2 to 3 hours after setting.
○ *crater with a sharp rim and a hilly floor (20 km/2 530 m)*

Small craters: Draper C (7.8 km/1 610 m)
Pytheas G (3.4 km/490 m)
Pytheas D (5.2 km/370 m)
Pytheas A (6.0 km/1 180 m)

PYTHEAS

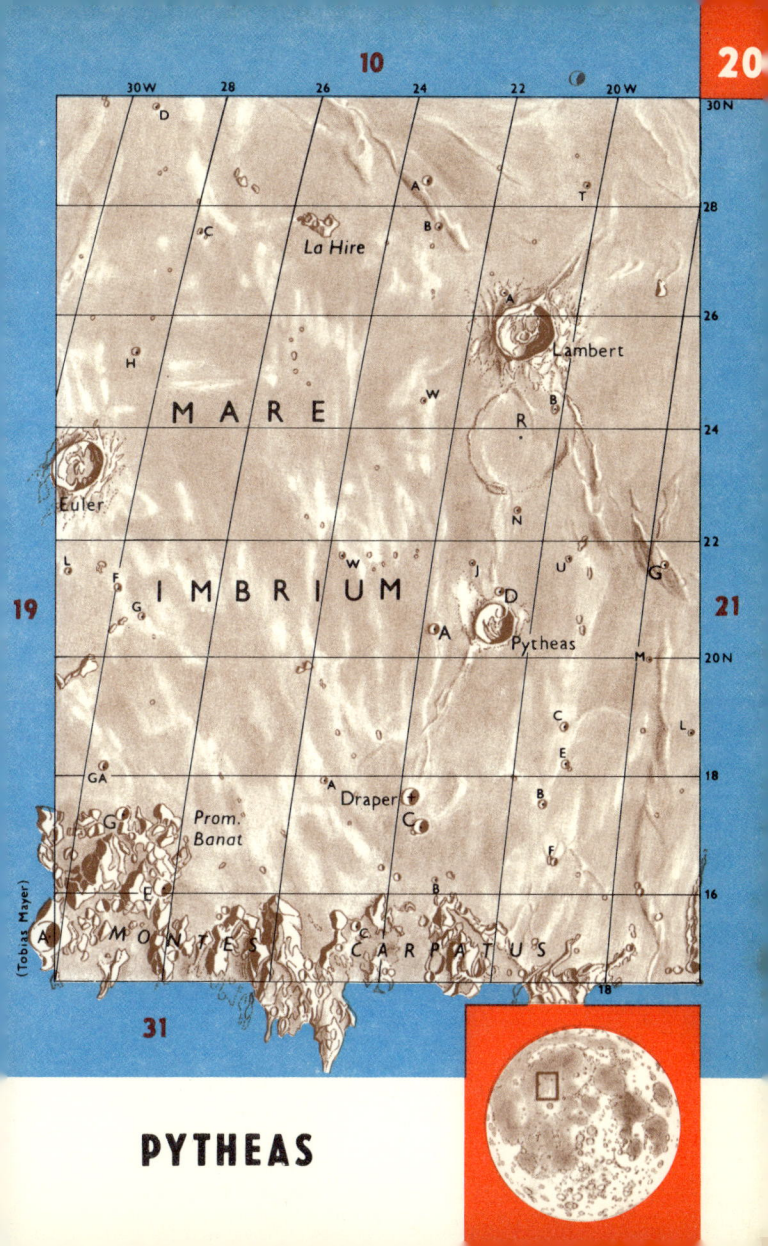

- La Hire
- Lambert
- MARE IMBRIUM
- Euler
- Pytheas
- Draper
- Prom. Banat
- MONTES CARPATUS
- (Tobias Mayer)

21 TIMOCHARIS

The south-eastern part of Mare Imbrium and south-western promontory of Apennines, terminated by crater Eratosthenes. Bright rays radiating from crater Copernicus (see map 31) are visible in oblique light on the surface of Mare Imbrium. A system of wrinkle ridges and a small dome can be seen in oblique light about 15 km south-south-east of crater Beer.

APENNINUS, MONTES Apennines, see map 22.
BEER (9W, 27N) Wilhelm Beer, 1797—1850; German selenographer; collaborator of Mädler, with whom he published a map of the Moon and a monograph 'Der Mond' (1837).
○ *crater with a sharp rim (10.2 km/1 650 m)*
ERATOSTHENES (11 W, 14 N) About 276—196 BC; Greek scholar, geographer and astronomer; first to measure the circumference of the Earth.
○ *very prominent crater with rather large terrace-like walls and central peaks (58 km/3 570 m)*
FEUILLÉE (10W, 27N) Louis Feuillée, 1660—1732; French; member of Franciscan order; botanist; Director of Observatory at Marseilles for 21 years.
○ *crater with a sharp rim (10 km/1 810 m)*
IMBRIUM, MARE Sea of Rains, see map 11.
TIMOCHARIS (13W, 27N) About 280 BC; Greek astronomer of the Alexandrian school; his measurements of the angular distance of the star Spica from the autumnal equinox enabled Hipparchus to discover precession 150 years later.
○ *prominent crater with a sharp rim and terrace-like walls (35 km/3 110 m)*
WALLACE (9W, 20N) Alfred Russel Wallace, 1823—1913; English naturalist and explorer; independently of Darwin he postulated a theory of the origin of species through natural selection. ○ *remains of flooded crater (28 km)*
WOLFF (MONS) (7W, 17N) Christian Freiherr von Wolff, 1679—1754; German philosopher and mathematician.
○ *mountain massif in the south-west promontory of the Apennines*
Small craters: Archimedes A (13.1 km/2 490 m)
Archimedes F (7.5 km/360 m)
Timocharis A (7.4 km/1 420 m)
Timocharis AA (2.8 km/400 m)
Timocharis B (5.2 km/940 m)

TIMOCHARIS

22 CONON

The largest lunar mountain chain of the Apennines and the prominent crater Archimedes dominate this part of the Moon close to the prime meridian.

AMPÈRE (MONS) (4W, 19N) André Marie Ampère, 1775—1836; French physicist.
○ *mountain massif in the central part of Apennines.*

APENNINUS, MONTES Apennine Mountains. Name given by Hevelius to the largest lunar mountain range on the south-east border of Mare Imbrium. This forms a part of the mare wall belonging to Mare Imbrium, into which it descends relatively steeply (about 30°). Slopes of Apennines towards Mare Vaporum are gradual. The height of the range exceeds 5 000 m in some places.

ARATUS (4E, 24N) About 315—245 BC; popular Greek poet; author of the oldest description of some 48 constellations.
○ *crater (10.6 km/1 860 m)*

ARCHIMEDES (4W, 30N) About 287—212 BC; Greek mathematician of Syracuse.
○ *very prominent flooded crater with terraces (83 km/2 150 m)*

BRADLEY (MONS) (0, 22 N) James Bradley, 1692—1762; English astronomer; discovered the aberration of light, and the effect known as nutation.
○ *mountainous massif in the vicinity of crater Conon*

CONON (2E, 22N) About 260 BC; Greek mathematician and astronomer, friend of Archimedes.
○ *prominent crater with a sharp rim (22 km/2 320 m)*

FRESNEL, PROM. (5E, 29N) Augustin Jean Fresnel, 1788—1827; French physicist, distinguished in optics (Fresnel's lens).
○ *Cape Fresnel, the northern promontory of Apennines*

HADLEY (MONS) (5E, 27N) John Hadley, 1682—1743; English physicist; made a number of astronomical instruments.
○ *mountainous massif in the northern part of Apennines*

HUYGENS (MONS) (3W, 20N) Christiaan Huygens, 1629—1695; Dutch astronomer, optician, mechanic.
○ *mountainous massif in the central part of Apennines; its height is 5 400 m*

MARCO POLO (2W, 15N) 1254—1324; famous Venetian traveller; travelled in Central Asia, China, Mongolia, Java, Sumatra, Madagascar and elsewhere.
○ *remains of an elongated crater (28 × 21 km)*

PUTREDINIS, PALUS Named by Riccioli.
Galen, Hadley, Huxley, Spurr, Yangel, see pp. 242, 243.

CONON

Map 22

- Archimedes
- Autolycus
- Pr. Fresnel
- Rimae Fresnel
- PALUS PUTREDINIS
- Spurr
- Hadley
- Apollo 15
- Hadley
- Bradley
- Rima Bradley
- APENNINUS
- Aratus
- Conon
- Galen
- Huxley
- Ampère
- Huygens
- MONTES
- Yangel
- Marco Polo
- MARE VAPORUM

23 LINNÉ

The western part of Mare Serenitatis offers the observer not only crater pits and long wrinkle ridges, but also the 'mysterious' crater Linné, which has been written about a great deal in the past. There are some clefts along the crater Sulpicius Gallus, which follow the edge of the mountain range Haemus.

HAEMUS, MONTES Haemus Mountains. An old name for a Balkan mountain range, originally given by Hevelius ('Haemus, mons Thraciae').

LINNÉ (12E, 28N) Carl von Linné (Linnaeus), 1707—78; Swedish botanist, physician, traveller; author of a system of nomenclature for plants and animals, founder of systematic botany.
  ○ *small crater with a sharp rim (2.4 km/600 m). The crater is surrounded by light material and under illumination, when the Sun's altitude is high, it resembles a brightly shining white patch. In the past numerous observers have recorded changes in the size and appearance of this formation.*

MANILIUS (9E, 14N) 1st century BC; Roman poet; author of the poem 'Astronomicon' which contains a description of long-known constellations. ○ *very prominent crater with terraces and central peaks (39 km/3 050 m)*

MENELAUS (16E, 16N) About 100 AD; Greek geometer and astronomer of Alexandria; author of 'Spherica', which deals with spherical trigonometry. ○ *prominent crater with a sharp rim and central peaks (27 km/3 010 m)*

SERENITATIS, MARE Sea of Serenity, see map 24.

SULPICIUS GALLUS (12E, 20N) About 168 BC; Roman consul, orator and scholar; in 168 BC, before the Battle of Pydna, he warned the Roman soldiers of the forthcoming eclipse of the Moon and thus helped them to remain calm whilst the enemy was overcome by panic.
  ○ *circular crater with a sharp rim (12.2 km/2 160 m)*

Small craters: Bessel E (6.5 km/1 230 m)
  Linné A (4.2 km/800 m)
  Bessel G (1 km/70 m)
  Bessel F (0.5 km/50 m)

Craters Bessel F and G appear as mere bright patches when viewed from the Earth. Their size was determined from the photographs taken by Lunar Orbiter 4.

Banting, Bowen, Daubrée, Hornsby, Joy, see pp. 241, 242.

23

MARE SERENITATIS

- Linné
- A
- D
- Banting
- Joy
- D
- C
- Hornsby
- (Aratus)
- Bessel
- A
- H
- F
- G
- G
- Sulpicius Gallus
- E
- MONTES HAEMUS
- E
- Bowen
- B
- A
- Menelaus
- B
- Z
- Daubrée
- G
- N
- C
- Manilius

LINNÉ

24 BESSEL

The eastern part of Mare Serenitatis is in many places intersected by wrinkle ridges, the largest of which approximately follows the line of longitude 25 E and has a zigzag shape (Serpentine Ridge). The edge of the Sea of Serenity is dark and has a number of shallow rilles. The summit of Posidonius Gamma has a crater pit 2 km in diameter.

ARCHERUSIA, PROM. Old name for a cape on the southern shores of Pontus Euxinus (Black Sea), which on Hevelius' map was marked in place of Mare Serenitatis and Mare Tranquillitatis.

ARGAEUS (MONS) Old name for a mountain in Cappadocia.

AUWERS (17E, 15N) Georg Friedrich Julius Arthur von Auwers, 1838—1915; German astronomer, Director of Observatory at Gotha. ○ *flooded crater, open to the north (20 km/1 680 m)*

BESSEL (18E, 22N) Friedrich Wilhelm Bessel, 1784—1846; German astronomer; one of the first to measure the parallax of a star (61 Cygni) in 1838.
○ *prominent crater (16 km/1 740 m)*

DAWES (26E, 17N) William Rutter Dawes, 1799—1868; English physician and amateur astronomer — excellent observer of planets, the Sun and double stars.
○ *crater with a sharp rim (18 km/2 330 m)*

DESEILLIGNY (20E, 21N) Jules A. P. Deseilligny, 1868—1918; French selenographer. ○ *crater (6.6 km/1 190 m)*

PLINIUS (24E, 15N) Pliny, about 23—79 AD; author of 'Historia Naturalis' in 37 books; died during the destruction of Pompeii.
○ *prominent crater with a sharp lip, terraces and central peaks (43 km/2 320 m)*

SERENITATIS, MARE Sea of Serenity. Circular mare with a surface area of 318 000 sq km. It is smaller than the Caspian Sea on Earth, which measures 371 000 sq km.

TACQUET (19E, 17N) André Tacquet, 1612—60; Belgian mathematician. ○ *crater (6.9 km/1 260 m)*

Small craters: Very (le Monnier B) (5.1 km/950 m)
Sarabhai (Bessel A) (7.6 km/1 660 m)
Bessel H (3.5 km/650 m)

Brackett, Sarabhai, Very, see pp. 242, 243.

BESSEL

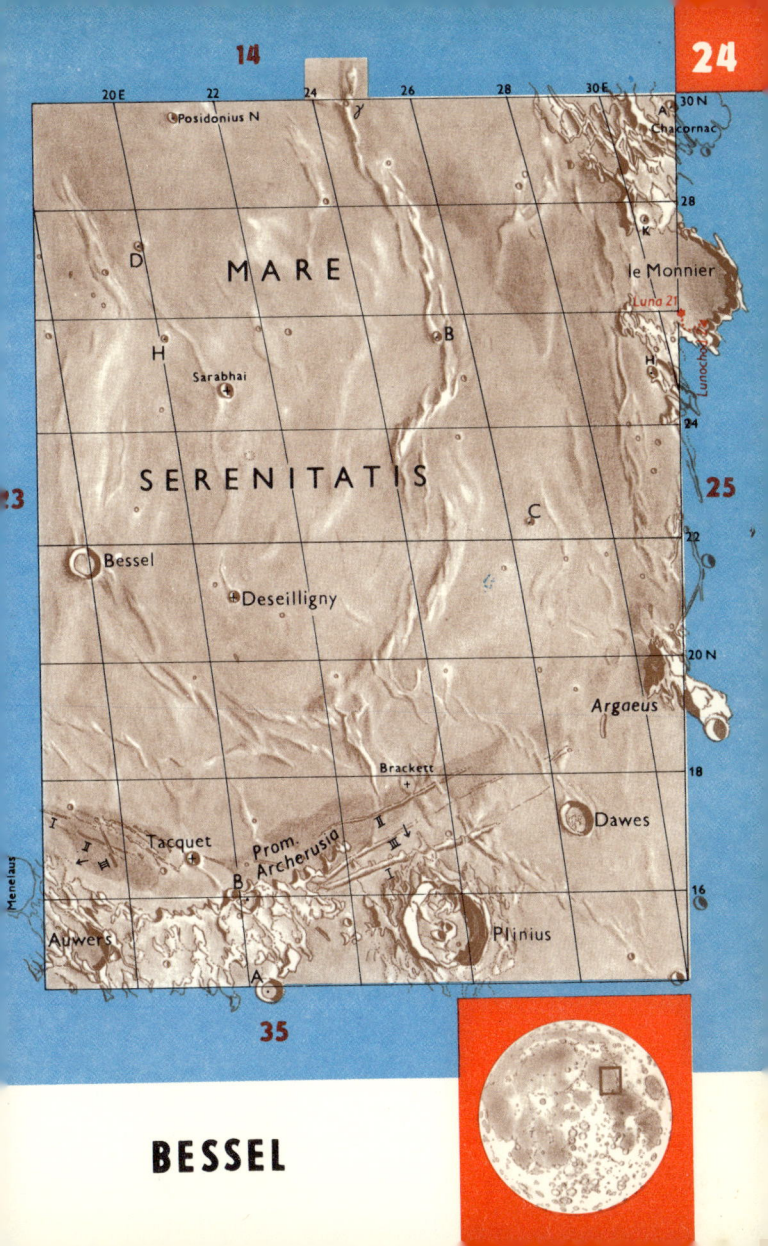

25 RÖMER

Continental mountainous area adjoining Sea of Serenity and Sea of Tranquillity. Crater Römer is surrounded by a dense field of craters with considerably disintegrated walls. The region of Römer, Chacornac and Littrow has a system of rilles visible even through smaller telescopes. Apollo 17 landed in the valley between the mountains situated to the north of crater Littrow. Lunokhod 2 was active in the crater le Monnier.

CHACORNAC, see map 14.

FRANZ (40E, 16N) Julius H. Franz, 1847—1913; German astronomer, selenographer.
○ *flooded, considerably eroded crater (26 km/590 m)*

LE MONNIER (30E, 26N) Pierre Charles le Monnier, 1715—99; French astronomer and physicist; measured degrees in Lapland.
○ *flooded crater with a very dark floor, which forms a small bay in the Sea of Serenity (61 km/2 400 m). See also map 24.*

LITTROW (31E, 22N) Joseph Johann von Littrow, 1781—1840; Austrian astronomer, director of Viennese observatory.
○ *flooded crater, with southern wall interrupted (31 km)*

MARALDI (35E, 19N) Giovanni Domenico Maraldi, 1709—88; Italian astronomer, assistant to Domenico Cassini.
○ *flooded crater with a very dark floor (40 km)*

NEWCOMB (44E, 30N) Simon Newcomb, 1835—1909; American mathematician and astronomer. ○ *crater (39 km/2 180 m)*

RÖMER (36E, 25N) Ole Rømer, 1644—1710; Danish astronomer; first to measure the speed of light by observing Jupiter's satellites. ○ *prominent crater with a sharp rim, terraces and a central peak (40 km)*

TAURUS, MONTES Mountain chain Taurus. Name given by Hevelius to the mountainous region north of Römer. At present the name is used for the inconspicuous mountain range between the craters Newcomb and Römer P.

VITRUVIUS (31E, 18N) Pollio Vitruvius, 1st century BC; Roman architect, author of 'De Architectura'; dealt also with astronomy, and sun and water clocks.
○ *flooded crater (28 km/1 550 m)*

Small craters: Jansen C (8.4 km/1 000 m)
Vitruvius E (11.1 km/2 090 m)
Vitruvius A (18.4 km/3 000 m)

Carmichael, Clerke, Franck, Hill, Lucian, Theophrastos, see pp. 242, 243.

RÖMER

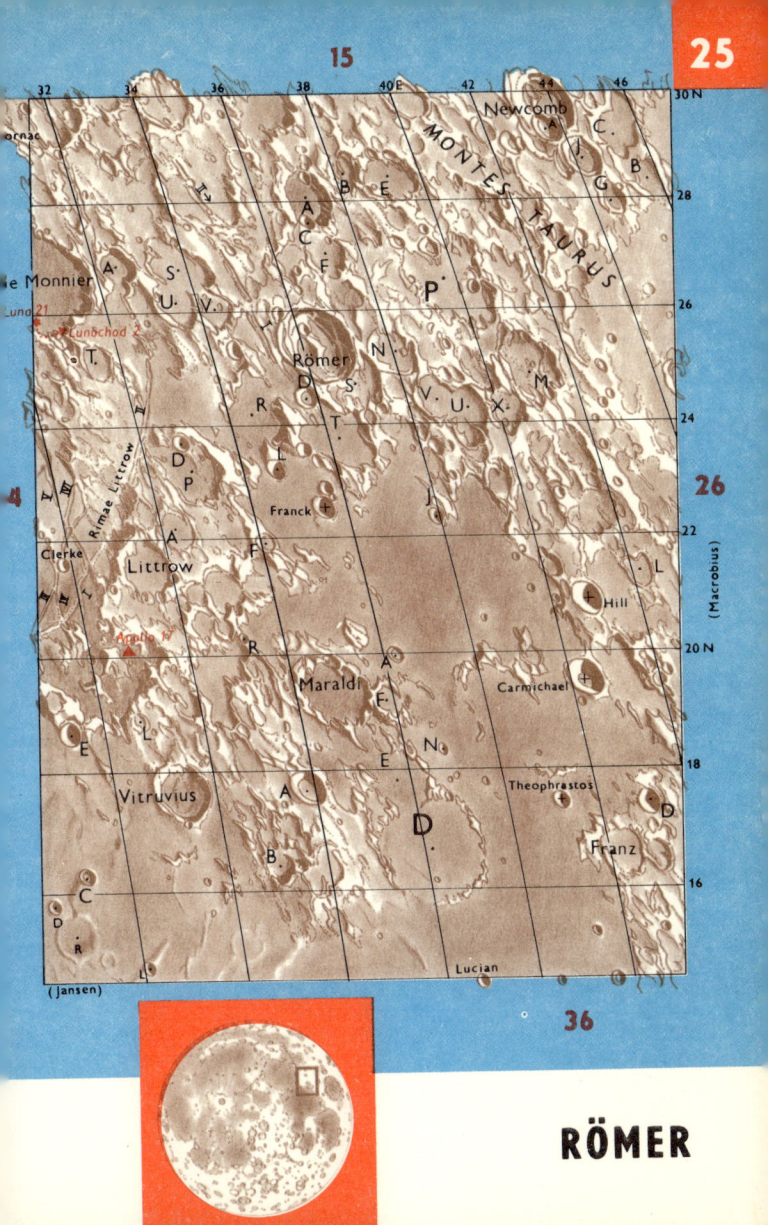

26 CLEOMEDES

The north-western part of Mare Crisium is encircled by large mountain massifs. Under oblique illumination, numerous wrinkle ridges are prominent on the surface of M. Crisium.

CLEOMEDES (56E, 28N) 1st century BC or later; Greek author of an astronomical work on Eratosthenes' measurements of the Earth, on the motions of the planets etc. ○ *very prominent crater (126 km) with clefts on the floor and a central peak.*
CRISIUM, MARE Sea of Crises, see map. 27.
DEBES (53E, 30N) Ernest Debes, 1840–1923; German cartographer; prepared lunar maps and atlases.
○ *crater (31 km) connected with crater Debes A*
DELMOTTE (60E, 27N) Gabriel Delmotte, 1876–1950; French selenographer. ○ *crater (33 km)*
LAVINIUM, PROM.
OLIVIUM, PROM. Two sharp 'Capes', situated opposite each other on the western edge of M. Crisium (named by Birt). These two promontories are separated by two crater pits and not a bridge, which was formerly claimed to be visible.
MACROBIUS (46E, 21N) Ambrosius A. T. Macrobius, 4th century AD; Greek grammarian, author of a commentary to Cicero's 'Scipion's Dream' (vision of the flight to the stars).
○ *prominent crater with terraces (64 km)*
PEIRCE (53E, 18N) Benjamin Peirce, 1809–80; American mathematician and astronomer. ○ *crater (18.5 km)*
PICARD (54E, 14N) Jean Picard, 1620–82; French astronomer, founder of the ephemeris 'Connaissance des Temps'.
○ *prominent crater with a sharp rim (23 km)*
PROCLUS (47E, 16N) 410–485 AD; Athenian philosopher.
○ *polygonal crater with sharp contours, a centre of radiating bright rays (28 km)*
SOMNI, PALUS Marsh of Sleep, see map 37.
TISSERAND (48E, 21N) François Félix Tisserand, 1845–96; French astronomer; worked in the field of celestial mechanics; established criteria for identifying the orbits of comets.
○ *crater (36 km)*
TRALLES (53E, 28N) Johann G. Tralles, 1763–1822; German physicist; invented an alcoholometer. ○ *crater (43 km)*
YERKES (53E, 14N) Charles T. Yerkes, 1837–1905; Chicago millionaire; financed observatory with the largest refractor in the world (opened in 1897). ○ *flooded crater (36 km)*
Curtis, Eckert, see p. 242.

CLEOMEDES

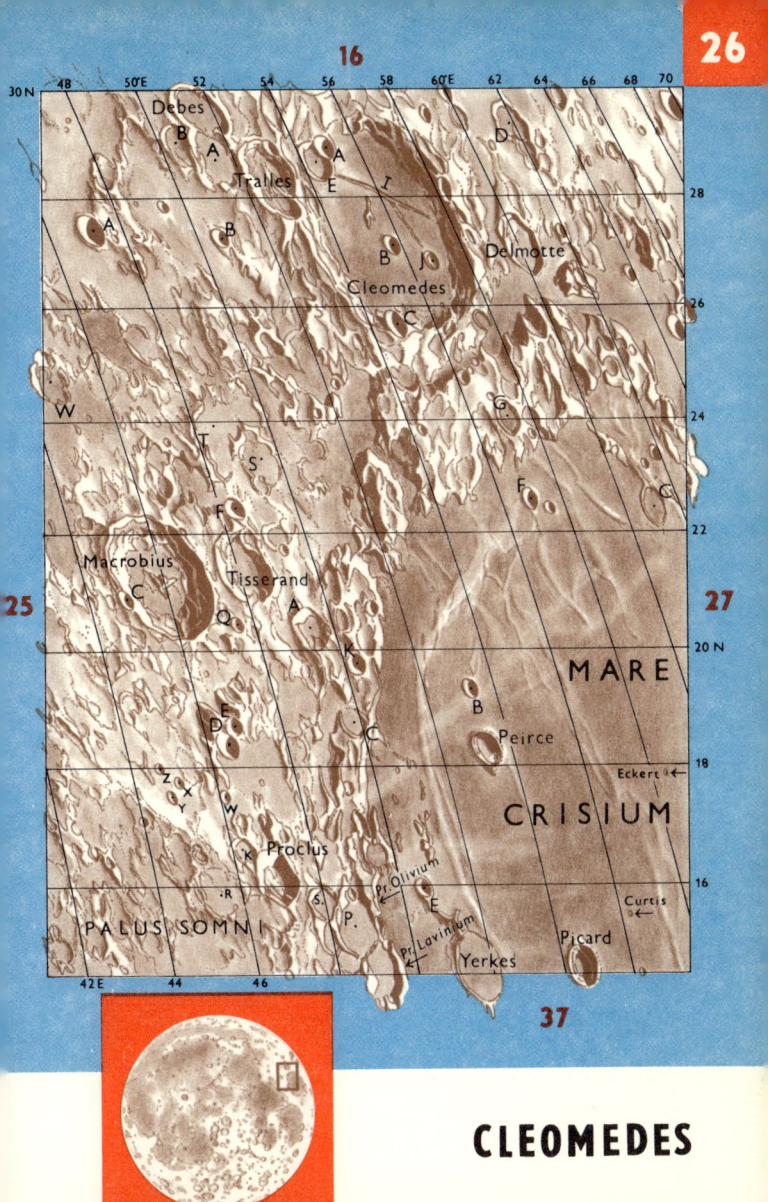

27 PLUTARCH

The eastern part of Mare Crisium and eastern margin of the Moon. Useful points of orientation are the craters Eimmart, Alhazen and Plutarch. During a favourable libration crater Goddard, with its dark floor, and Mare Marginis are clearly visible.

ALHAZEN (72E, 16N) Abu Ali al-Hasan, 987—1038; Arabian mathematician at the court of Caliph Hakem in Cairo.
○ *prominent crater with a sharp rim to its wall (33 km)*

ANGUIS, MARE The 'Serpent Sea'. Franz's name for a narrow, sinuous valley.

CANNON (20N, 81E) Annie Jump Cannon, 1863—1941; American astronomer; worked on the classification of stellar spectra. ○ *flooded crater with a lighter floor (57 km)*

CRISIUM, MARE The 'Sea of Crises'. Oval mare, elongated to east-west, surrounded by a mountainous wall. Its surface area (199 000 sq km) is comparable with that of Great Britain.

EIMMART (65E, 24N) Georg Christoph Eimmart, 1638—1705; German engraver and astronomer, made map of the Moon.
○ *crater (46 km)*

GODDARD (89E, 15N) Robert Hutchings Goddard, 1882—1945; American physicist, pioneered rocket technology.
○ *flooded crater, dark floor (85 km)*

HUBBLE (87E, 22N) Edwin Powell Hubble, 1889—1953; American astronomer; investigation into galaxies; discovery of 'Hubble's relationship'. ○ *partly flooded crater (82 km)*

LIAPUNOV (89E, 27N) Alexander Mikhailovich Liapunov, 1857—1918; Russian mathematician and engineer.
○ *crater (66 km)*

MARGINIS, MARE The 'Border Sea'. A small, irregularly shaped mare at the eastern edge of the Moon, named by Franz.

PLUTARCH (79E, 24N) About 46—120 AD; Greek biographer. In his dialogue 'De facie in orbe lunae (Concerning the Face which appears in the Orb of the Moon)' he developed some earlier theories on the nature of the Moon.
○ *prominent crater (68 km)*

RAYLEIGH (90E, 29N) John William Strutt Baron Rayleigh, 1842—1919; English physicist; awarded Nobel Prize in 1904; research in optics. ○ *walled plain (102 km) along the edge*

SENECA (80E, 27N) Lucius A. Seneca, 3 BC — 65 AD; Roman statesman and orator, teacher of Nero; in his work 'Quaestiones Naturales' he concluded that comets are celestial bodies.
○ *inconspicuous, irregularly shaped crater (63 km)*

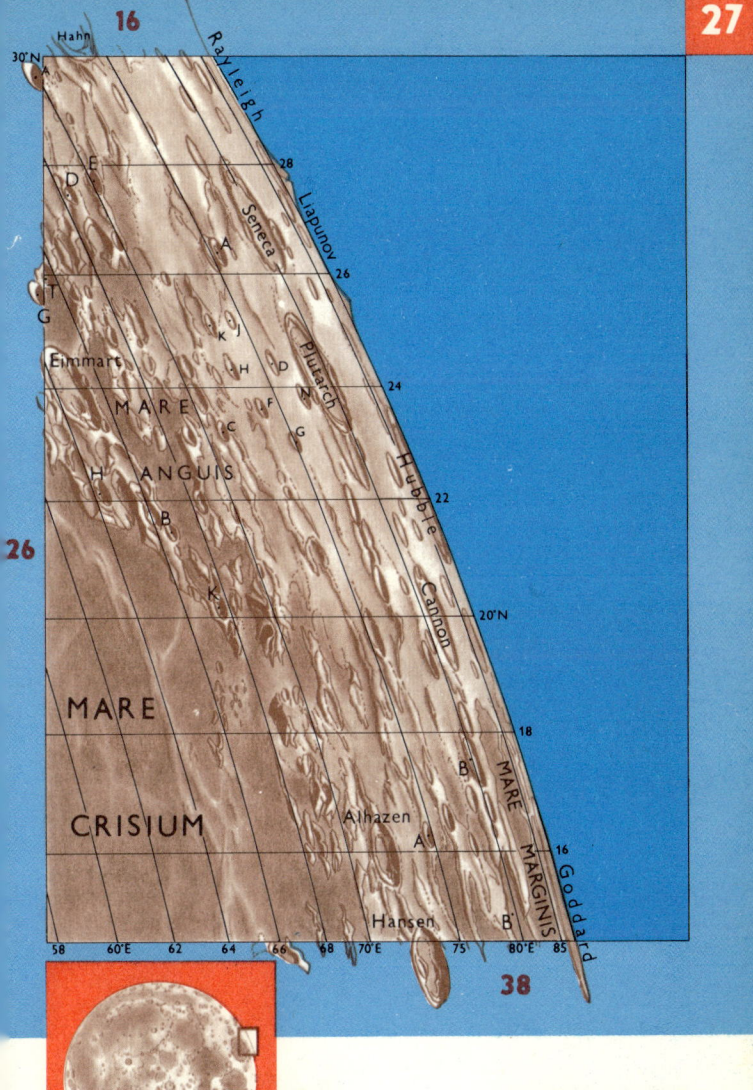

PLUTARCH

28 GALILAEI

The western margin of the Moon and western edge of Oceanus Procellarum. The observer using a large telescope will be attracted by the clefts in the floor of Hevelius and Rima Cardanus and the unique formation Reiner Gamma. Landing area of Luna 8 and 9 (the first soft landing on the Moon).

BOHR (87W, 13N) Niels Henrik David Bohr, 1885–1962; Danish physicist; best known for his model of the atom; Nobel Prize in 1922. ○ *crater in the libration zone (73 km)*

CARDANUS (72W, 13N) Girolamo Cardano, 1501–76; Italian mathematician, astrologer and physician.
○ *crater (50 km), connected by a cleft with Krafft (see map 17).*

CAVALERIUS (67W, 5N) Buonaventura Cavalieri, 1598–1647; Italian mathematician, pupil of Galilei.
○ *prominent crater (64 km)*

GALILAEI (63W, 10N) Galileo Galilei, 1564–1642; Italian natural philosopher and astronomer; first telescopic observations of the sky; advocate of Copernican theories; observed sunspots, the Moon, discovered Jupiter's satellites.
○ *crater with a sharp rim (15.5 km)*

HEDIN (76W, 3N) Sven Anders Hedin, 1865–1952; Swedish explorer; made expeditions to Central Asia.
○ *remains of a walled plain (143 km)*

HEVELIUS (68W, 2N) Johann Hewelcke (Hevel), 1611–87; Danzig astronomer, selenographer; proposed a new terminology for the Moon, of which only six names have been preserved.
○ *walled plain (118 km)*

OLBERS (76W, 7N) Heinrich Wilhelm Matthäus Olbers, 1758–1840; German physician and astronomer; discovered and observed comets; 'Olbers paradox'. ○ *crater (71 km)*

PLANITIA DESCENSUS Plain of Descent (64W, 7N). Place of the first soft landing on the Moon made by Luna 9. It is situated between two small hills at the edge of the Ocean of Storms.

PROCELLARUM, OCEANUS Ocean of Storms, see map 29.

REINER GAMMA (crater Reiner, see map 29). Entirely flat feature, formed by very light material.

VASCO DA GAMA (83W, 14N) 1469–1524; Portuguese navigator; made the first voyage to India round the Cape of Good Hope (1498). ○ *crater (90 km)*

GALILAEI

29 MARIUS

The western part of Oceanus Procellarum, poor in large craters, but very rich in lunar domes, especially near crater Marius. This area, geologically very interesting, with distinct traces of volcanic activity, was chosen as the target for one of the Apollo missions, but the mission did not materialize because the programme was curtailed.

MAESTLIN (41W, 5N) Michael Möstlin (Maestlin), 1550–1631; German mathematician and astronomer, teacher of Johannes Kepler, whom he introduced to the Copernican heliocentric system.
○ *small crater (7.1 km/1 650 m), located to the north of the remains of the walled plain Maestlin R (60 km)*

MARIUS (51W, 12N) Simon Mayer, 1570–1624; German astronomer, independently discovered Jupiter's satellites.
○ *regular flooded crater (41 km). A small crater Marius G (3.3 km) rises from its flat floor.*

PROCELLARUM, OCEANUS Ocean of Storms. The largest lunar mare. Its surface area is 2 100 000 sq km – i.e., 866 000 sq km smaller than the Mediterranean on Earth. The western, northern and southern borders of the Ocean are relatively distinct, while the eastern edge is indefinite. The surface is furrowed by numerous wrinkle ridges. The laser altimeter of Apollo 15 discovered some exceptionally flat places in this Ocean with height differences ranging between ± 80 m over a total area of 200 km. The bright rays radiating from crater Kepler are visible in this area.

REINER (55W, 7N) Vincentio Reinieri, died 1648; Italian mathematician, pupil and friend of Galilei.
○ *prominent crater (30 km)*

SUESS (48W, 4N) Eduard Suess, 1831–1914; Austrian geologist and selenologist; advocated cosmic origin of tektites.
○ *small crater (9.2 km)*

Small craters: Kepler C (12.2 km/2 170 m)
Kepler D (10.0 km/350 m)
Kepler E (5.2 km/1 000 m)
Maestlin G (2.8 km/670 m)
Maestlin H (7.1 km/1 370 m)

29

OCEANUS

Marius
Luna 7

Reiner

Maestlin

Suess

PROCELLARUM

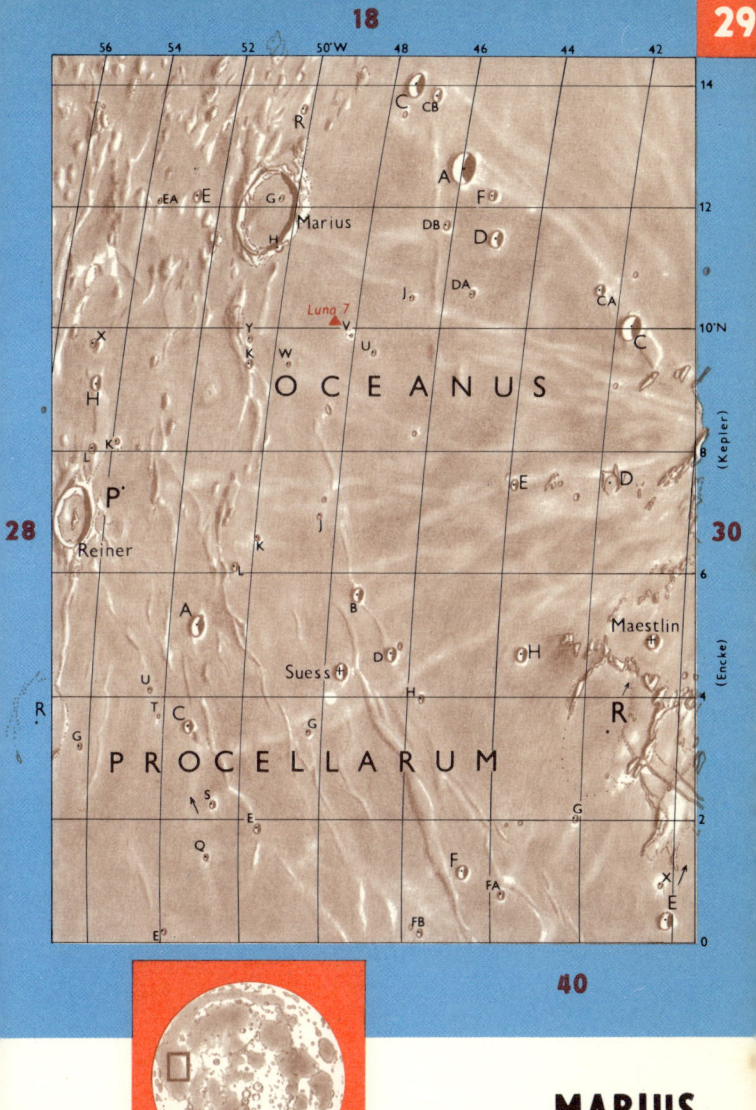

(Kepler)
(Encke)

MARIUS

30 KEPLER

The dark background of Oceanus Procellarum is a focus of bright rays radiating from two main centres — crater Kepler, which dominates this area, and crater Copernicus (in the east). To the north of crater Hortensius is situated perhaps the best-known group of lunar domes. Other domes can be seen even through a small telescope to the west of Milichius and in the neighbourhood of mountain massifs of Tobias Mayer Alfa and Dzeta.

ENCKE (37W, 5N) Johann Franz Encke, 1791—1865; German astronomer who calculated the elements of the orbit of 'Encke's comet' (the first known short-periodic comet).
○ *crater with uneven floor (29 km/750 m). Small crater Encke N (3.5 km/590 m) is situated on its western wall.*

HORTENSIUS (28W, 6N) Martin van den Hove, 1605—39; Dutch astronomer.
○ *small crater with a sharp rim (14.6 km/2 860 m). To the north there is a group of domes, most of them having peak craters.*

KEPLER (38W, 8N) Johannes Kepler, 1571—1630; German astronomer, ingenious theoretician, who on the basis of Tycho Brahe's observations formulated three laws relating to the motions of planets around the Sun.
○ *very prominent crater (32 km/2 750 m) with uneven floor. Centre of radiating bright rays.*

KUNOWSKY (32W, 3N) Georg K. F. Kunowsky, 1786—1846; German lawyer and amateur astronomer, observer of the Moon and the planets.   ○ *flooded crater (18 km/850 m)*

MILICHIUS (3W, 10N) Jacob Milich, 1501—59; German physician, philosopher, mathematician.
○ *crater with a sharp rim (13 km/2 510 m). Milichius Pi is a typical dome with a peak crater.*

PROCELLARUM, OCEANUS Ocean of Storms, see map 29.
Small craters: Hortensius DB (5.9 km/200 m)
Hortensius B (6.7 km/1 170 m)
Hortensius A (10.2 km/1 850 m)
Encke B (11.5 km/2 230 m)

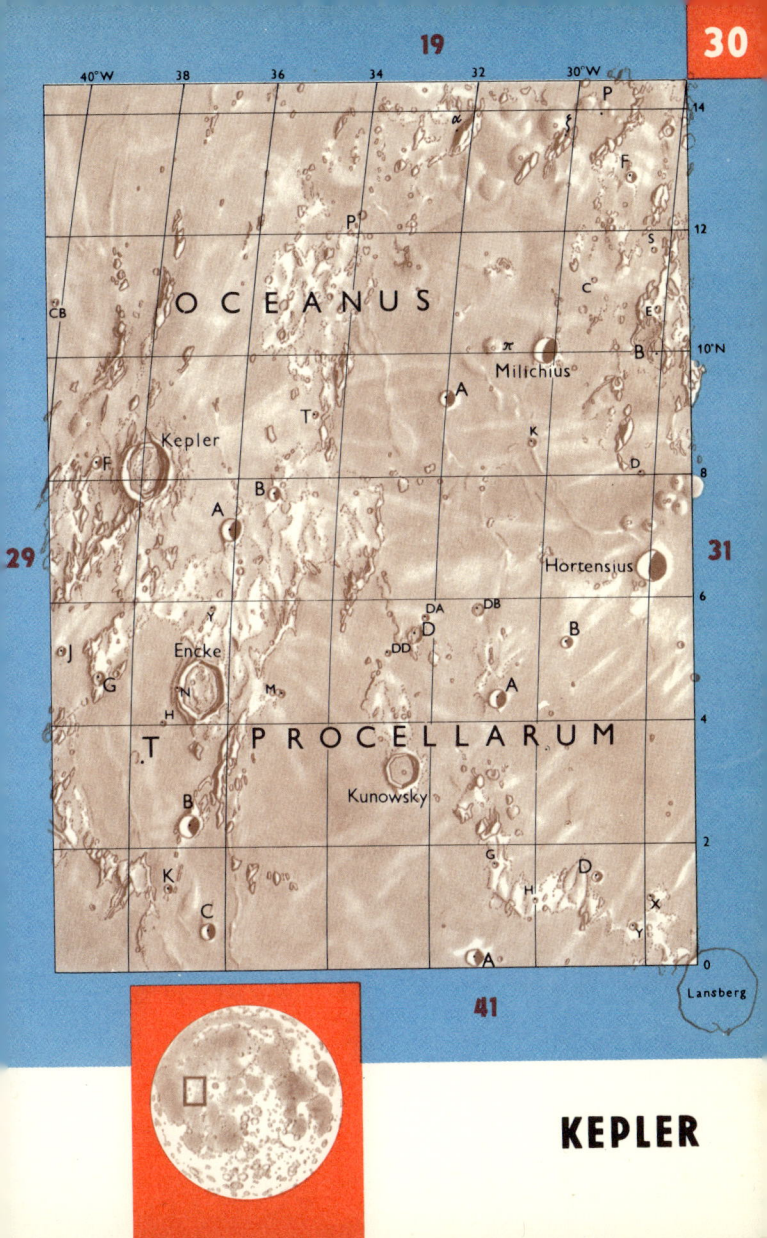

OCEANUS PROCELLARUM

Kepler
Milichius
Hortensius
Encke
Kunowsky
Lansberg

KEPLER

31 COPERNICUS

Crater Copernicus is undoubtedly one of the best-known and most typical lunar formations. It is a centre of bright rays, which can best be traced on the surface of Mare Imbrium. To the north of Copernicus is a group of scattered solitary hills, which rise to a height of several hundred metres.

CARPATUS, MONTES Carpathian Mountains. Mädler's name for a mountain range at the southern edge of Mare Imbrium. It stretches approximately east to west and its length is about 360 km. It consists of individual hills and mountain massifs, whose height fluctuates between 1 000 m and 2 000 m (see also maps 19 and 20).

COPERNICUS (20W, 10N) Nicholaus Copernicus, 1473–1543; Polish astronomer, one of the founders of modern astronomy; his heliocentric system is explained in his main work 'De Revolutionibus Orbium Celestium'.
○ *ring mountains 93 km in diameter and 3 760 m deep. Terrace-like walls and a relatively flat floor with a group of central peaks, whose height reaches 1 200 m. The height of the ridge of the wall above the surrounding landscape is 900 m. This formation is very suitable for observation by amateurs, even with a small telescope.*

FAUTH (20W, 6N) Philipp Johann Heinrich Fauth, 1867–1943; German selenographer and observer of planets; made lunar maps.
○ *double crater; Fauth and Fauth A resemble in shape a keyhole. Crater Fauth has a diameter of 12.1 km and a depth of 1 960 m; Fauth A is 9.6 km in diameter and is 1 540 m deep.*

GAY LUSSAC (21W, 14N) Joseph Louis Gay-Lussac, 1778–1850; French physicist and chemist; formulated a law of the expansion of gases by heat at constant pressure, later known as Charles' Law.   ○ *crater at the southern edge of the Carpathians, slightly disintegrated wall (26 km/830 m)*

REINHOLD (23W, 3N) Erasmus Reinhold, 1511–53; German mathematician and astronomer; friend of Rheticus; author of 'Prutdentine Tables' based on teaching of Copernicus.
○ *prominent crater, terraces (48 km/3 260 m)*

Small craters: Copernicus H (4.6 km/870 m) dark 'halo'
Gambart A (12 km/2 440 m)
Gay Lussac A (14 km/2 550 m)
Tobias Mayer C (15.6 km/2 510 m)
Tobias Mayer D (8.6 km/1 470 m)

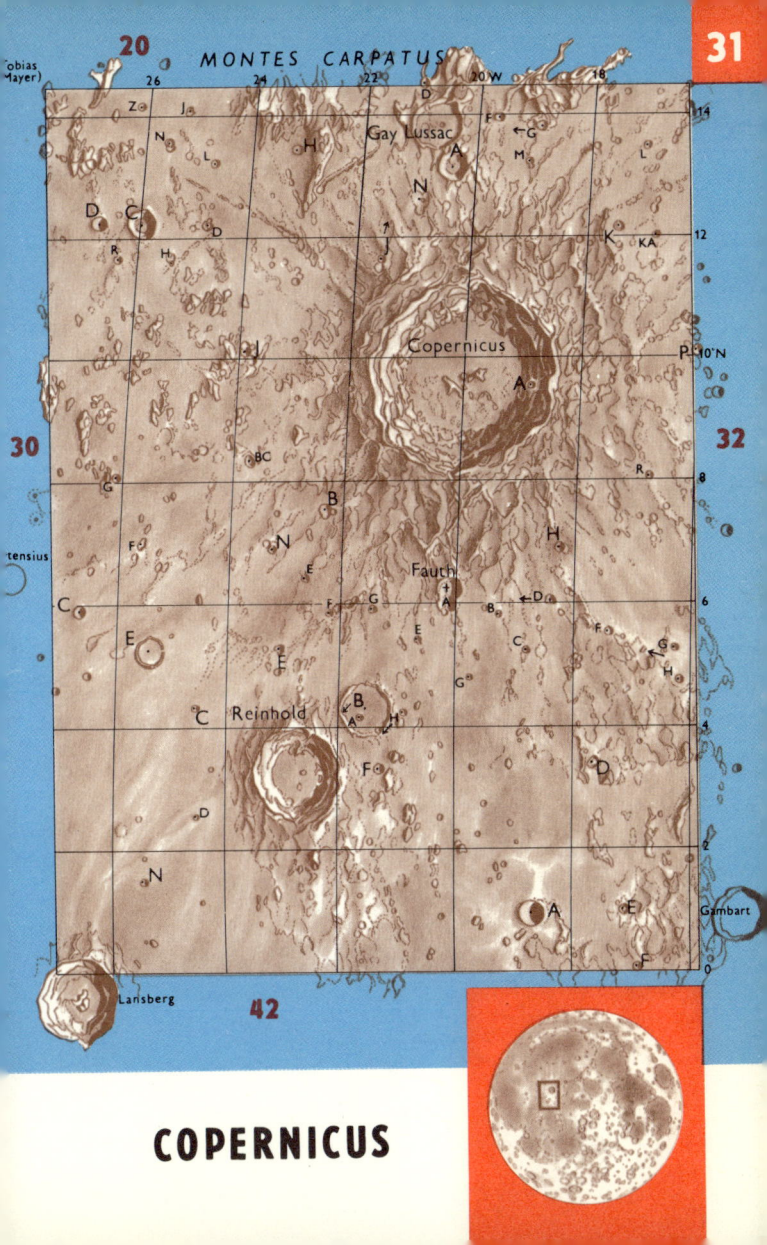

COPERNICUS

20 MONTES CARPATUS **31**

Gay Lussac

Copernicus

Fauth

Reinhold

Lansberg

Tobias Mayer

tensius

Gambart

30 **32**

42

32 STADIUS

The dark, monotonous area of Sinus Aestuum is surrounded by several interesting formations, the beautiful crater Eratosthenes being the most striking. A mountain range projects south-west of Eratosthenes towards the 'submerged crater' Stadius. North-west of Stadius rises a crater chain which continues further north (see map 20). A dome is clearly visible close to crater Gambart C. Under oblique illumination large prominent dark patches can be seen to the south of crater Copernicus C and to the north of crater Schröter D.

AESTUUM, SINUS Bay of Billows. Very flat area similar to mare, partly disintegrated by small, inconspicuous wrinkle ridges and crater pits.

ERATOSTHENES, see map 21. Together with the nearby remains of crater Stadius, this area offers an example of two completely different types of lunar craters, which vary both in appearance and origin. During the Full Moon, Eratosthenes almost disappears and is as faint as Stadius.

GAMBART (15W, 1N) Jean Felix Adolfe Gambart, 1800–36; French astronomer; discovered 13 comets.
  ○ *flooded crater with single wall (25.6 km/1 050 m)*

SCHRÖTER (7W, 3N) Johann Hieronymus Schröter, 1745–1816; German selenographer; experienced observer, author of 'Selenotopographische Fragmente'; discovered clefts and rilles on the Moon.
  ○ *considerably disintegrated wall, open in the south (34.5 km)*

SÖMMERING (8 W, 0.1 N) Samuel Thomas Sömmering, 1755–1830; German surgeon and naturalist.
  ○ *crater with considerably disintegrated wall (30 km)*

STADIUS (14W, 10N) Jan Stade, 1527–79; Belgian mathematician and astronomer; author of planetary tables 'Tabulae Bergenses'.
  ○ *ring depression, edged sporadically with a low wall and crater pits. Its diameter is 64 km and the height of the wall in the north-east is 650 m.*

Small craters: Gambart C (12.2 km/2 300 m)
  Gambart B (11.5 km/2 170 m)
  Schröter A (4.2 km/620 m)
  Schröter W (10.1 km/610 m)

STADIUS

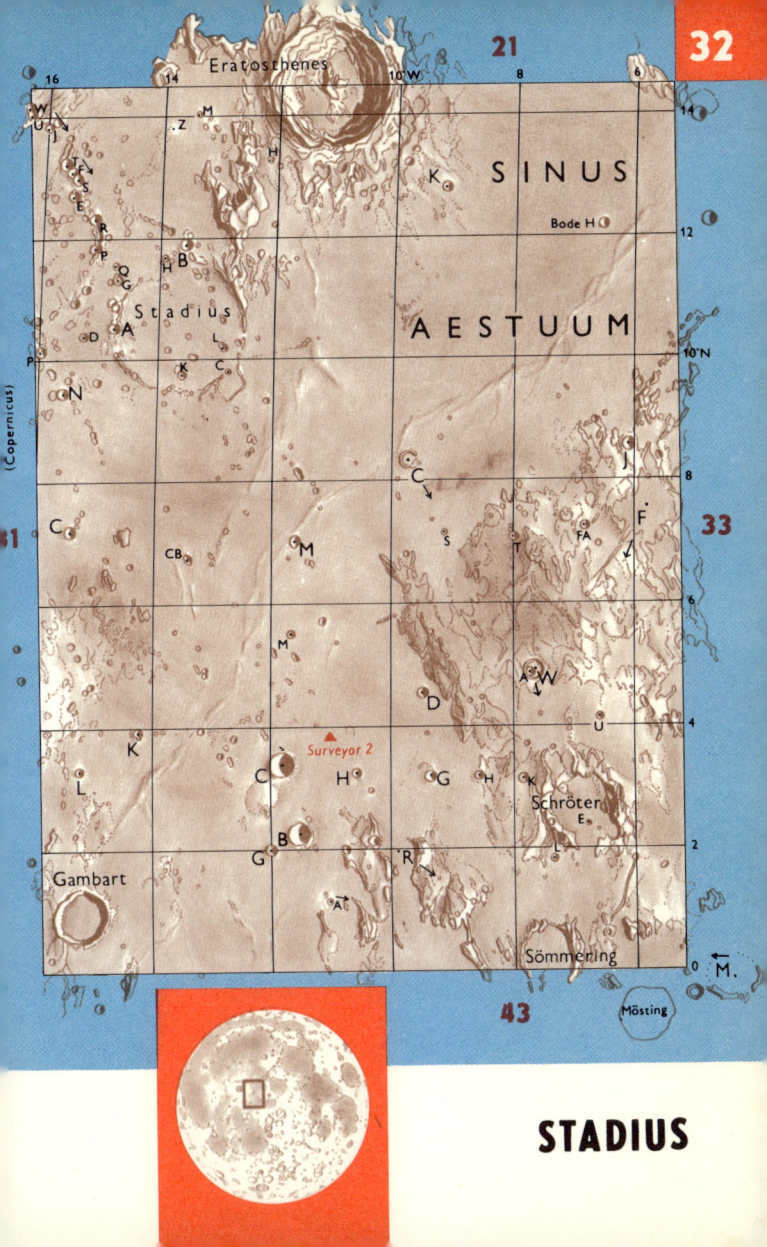

33 TRIESNECKER

A continental region, surrounded by the dark areas of Mare Vaporum, Sinus Medii and Sinus Aestuum. A complex system of clefts surrounds crater Triesnecker and a section of the Hyginus cleft is also visible.

AESTUUM, SINUS Bay of Billows, see map 32.

BLAGG (1E, 1N) Mary Adela Blagg, 1858—1944; English selenographer, who took an important part in preparing the modern lunar nomenclature accepted by IAU in 1935.
○ *small crater (5.4 km/910 m)*

BODE (2W, 7N) Johann Elert Bode, 1747—1826; German astronomer; commemorated in the Titius-Bode Law of planetary distances. ○ *crater (18.6 km/3 480 m)*

BRUCE (0.4E, 1.1N) Catherine W. Bruce, 1816—1900; American patron of art and science; supported a number of astronomers at home and abroad. ○ *small crater (6.7 km/1 260 m)*

CHLADNI (1.1E, 4N) Ernst Florence Friedrich Chladni, 1756—1827; German physicist; in 1794 he was first to demonstrate that meteors are of cosmic origin; 'Chladni's figures'.
○ *crater (13.6 km/2 630 m)*

MEDII, SINUS Central Bay. Mädler's name for a small mare at the centre of the near side of the Moon.

MURCHISON (0.1W, 5N) Sir Roderick Impey Murchison, 1792—1871; Scottish by origin, geologist and geographer.
○ *crater with a considerably disintegrated wall (58 km)*

PALLAS (2W, 6N) Peter Simon Pallas, 1741—1811; German naturalist and explorer; discovered the Pallas meteorite close to Krasnoyarsk. ○ *crater (50 km/1 260 m)*

RHAETICUS (5E, 0N) Georg Joachim Rheticus, 1514—76; German mathematician and astronomer, pupil of Copernicus.
○ *irregular crater with a disintegrated wall (43 × 49 km)*

TRIESNECKER (4E, 4N) Franz von Paula Triesnecker, 1745—1817; Austrian mathematician and astronomer; Director of Observatory at Vienna.
○ *prominent crater with central peaks (26 km/2 760 m)*

UKERT (1E, 8N) Friedrich August Ukert, 1780—1851; German historian and philologist. ○ *crater (24 km/2 800 m)*

VAPORUM, MARE Sea of Vapours. Riccioli's name for a circular mare situated to the south-east of the Apennines.

Small craters: Bode A (12.3 km/2 820 m)
Bode B (10.2 km/1 780 m)
Bode C (7 km/1 300 m)
Pallas A (10.6 km/2 080 m)

TRIESNECKER

34 HYGINUS

A region with a prominent radial structure, directed towards the basin of Mare Imbrium, containing rilles Rima Hyginus and Rima Ariadaeus, which can be observed easily. An interesting plateau is situated about 10 km north of the crater Godin C.

AGRIPPA (10 E, 4 N) About 92 AD; Greek astronomer; in 92 AD observed occultation of the Pleiades by the Moon.
○ *crater (46 km/3 070 m)*

BOSCOVICH (11E, 10N) Ruggiero Giuseppe Boscovich, 1711–1787; mathematician, physicist and astronomer from Dubrovnik.
○ *crater with a considerably disintegrated wall (46 km/1 770 m)*

CAYLEY (15E, 4N) Arthur Cayley, 1821–95; English mathematician and astronomer. ○ *ring crater (14.3 km/3 130 m)*

D'ARREST (15E, 2N) Heinrich Louis d'Arrest, 1822–75; German astronomer; worked in the field of comets and asteroids.
○ *crater with a disintegrated wall (30 km/1 490 m)*

DEMBOWSKI (7E, 3N) Baron Ercole Dembowski, 1815–81; Italian astronomer; measured 20 000 positions of double stars.
○ *crater (29 km), wall is open in the east*

DE MORGAN (15E, 3N) Augustus de Morgan, 1806–71; English mathematician. ○ *crater (10 km/1 860 m)*

GODIN (10E, 2N) Louis Godin, 1704–60; French mathematician and geodesist. ○ *crater (35 km/3 200 m)*

HYGINUS (6E, 8N) Caius Julius Hyginus, 1st century AD; Spanish by origin, friend of Ovidius; described constellations and their mythology.
○ *crater (10.6 km/770 m). Rima Hyginus is a shallow valley, formed sporadically by a row of craters.*

JULIUS CAESAR (15E, 9N) About 102–44 BC; Roman statesman; remembered by Riccioli because of his reform of the calendar. ○ *wide wall, dark floor (90 km)*

MANILIUS, see map 23.

SILBERSCHLAG (12E, 6N) Johann E. Silberschlag, 1721–91; German theologian, author of one of the first calculations of a meteor's orbit. ○ *crater (13.4 km/2 530 m)*

TEMPEL (12E, 4N) Ernest Wilhelm Leberecht Tempel, 1821–1889; German astronomer; discovered six asteroids and many comets. ○ *crater with a disintegrated wall (48 km/1 250 m)*

VAPORUM, MARE Sea of Vapours, see map 33.

WHEWELL (14E, 4N) William Whewell, 1794–1866; English philosopher, historian of science. ○ *crater (14 km/2 260 m)*

HYGINUS

35 ARAGO

The western part of Mare Tranquillitatis. Prominent network of wrinkle ridges surrounds the formation Lamont. Large domes of Alfa and Beta Arago and a group of four small domes north-west of Alfa Arago.

ARAGO (21E, 6N) Dominique Francois Jean Arago, 1786−1853; French astronomer. ○ *crater (26 km)*
ARIADAEUS (17E, 4N) Arrhidaeus, died 317 BC; Macedonian king, whose name was entered in the Babylonean list of eclipses.
○ *crater (11.2 km/1 830 m)*
DIONYSIUS (17E, 3N) St Dionysius; according to Riccioli he observed a solar eclipse when Christ was crucified.
○ *crater (17.6 km), very bright during Full Moon*
LAMONT (23E, 5N) John Lamont, 1805−79; German astronomer, Scottish by origin. ○ *inconspicuous formation 88 km in diameter; its wall is outlined by wrinkle ridges.*
MACLEAR (20E, 10N) Sir Thomas Maclear, 1794−1879; Irish descent, astronomer. ○ *flooded crater (20 km/610 m)*
MANNERS (20E, 4N) Russell Henry Manners, 1800−70; English admiral, astronomer. ○ *crater (15 km/1 710 m)*
RITTER (2N, 19E) 1. Karl Ritter, 1779−1859, German geographer. 2. August Ritter, 1826−1908, German astrophysicist. ○ *crater (31 km) with an uneven floor*
ROSS (22 E, 12 N) 1. Sir James Clark Ross, 1800−62; English polar explorer; Ross' Sea. 2. Frank E. Ross, 1874−1966; American astronomer. ○ *crater (26 km), elongated shape*
SABINE (20E, 1N) Sir Edward Sabine, 1788−1883; astronomer, Irish by origin. ○ *crater (30 km)*
SCHMIDT (19E, 1N) 1. Johann Friedrich Julius Schmidt, 1825−84; German selenographer. 2. Bernhard Schmidt, 1879−1935; German optician. 3. Otto J. Schmidt, 1891−1956; Soviet mathematician, explorer.
○ *ring crater (11.4 km/2 300 m)*
SOSIGENES (18E, 9N) About 46 BC; Greek astronomer, reformed the calendar. ○ *crater (18 km/1 730 m)*
TRANQUILLITATIS, MARE Sea of Tranquillity, see map 36.
TRANQUILLITATIS, STATIO Base of Tranquillity. Landing place of Apollo 11.
ARMSTRONG, ALDRIN, COLLINS. American astronauts of Apollo 11. ○ *crater pits: Armstrong (4.6 km/670 m), previously Sabine E. Aldrin (3.4 km/600 m), previously Sabine B. Collins (2.4 km/560 m), previously Sabine D.*

ARAGO

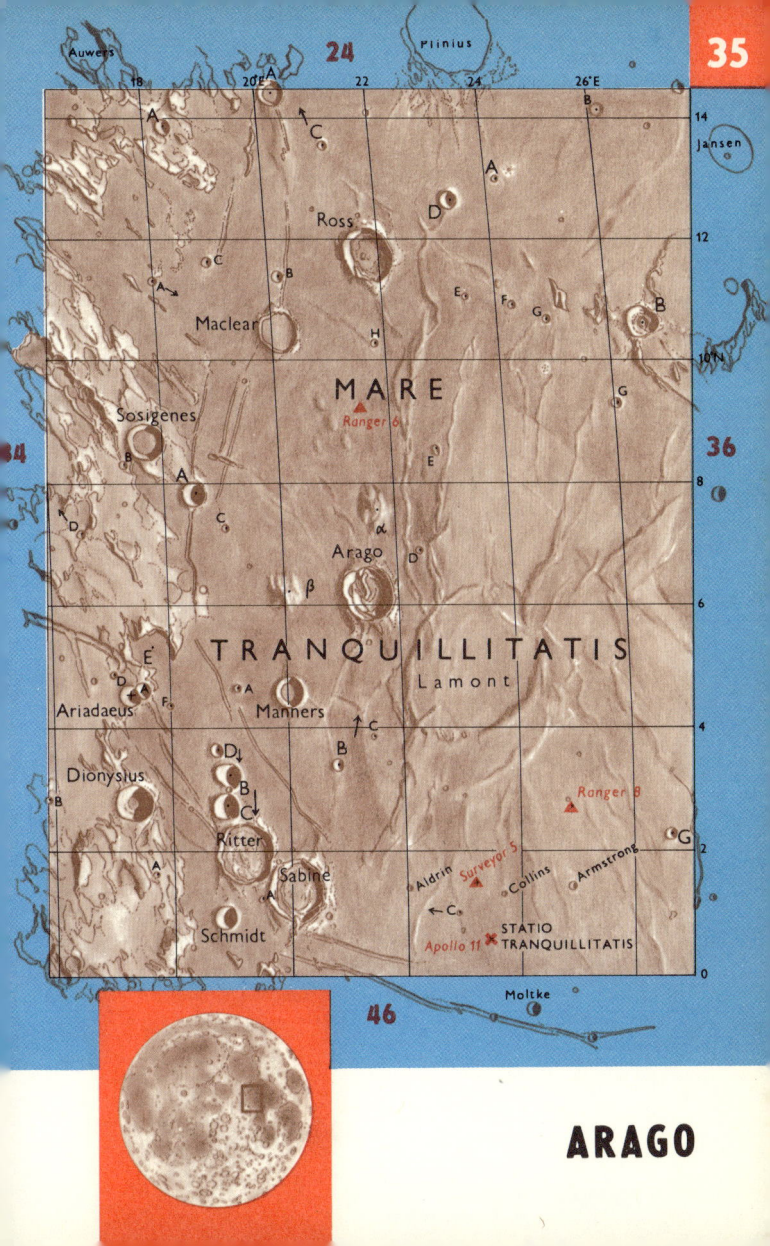

36 CAUCHY

The central part of Mare Tranquillitatis. The most interesting area is in the neighbourhood of crater Cauchy. Rilles are clearly visible here under suitable illumination near the terminator, even with a small telescope. There are also the striking fault Rupes Cauchy and the typical domes Omega and Tau. Large telescopes will reveal a peak crater on the dome Omega Cauchy. A narrow cleft along the edge of the map stretches west of crater Maskelyne.

CAJAL (previously Jansen F), see p. 243.
CAUCHY (39E, 10N) Augustin Louis Cauchy, 1789—1857; French mathematician.
 ○ *crater, very bright during Full Moon (12.4 km/2 610 m)*
CAUCHY, RUPES. A furrow narrowing into a cleft. During the sunrise the north-eastern wall casts a striking shadow but is bright in the setting sun. Similar formation Rupes Recta (see map 54).
JANSEN (29E, 14N) Zacharias Janszoon, died 1619; Dutch optician; credited with invention of telescope.
 ○ *flooded crater with a low wall (23 km/620 m)*
LYELL (41E, 14N) Sir Charles Lyell, 1797—1875; English geologist and explorer.
 ○ *crater with irregular, disintegrated wall and a dark floor (32 km)*
MASKELYNE (30E, 2N) Nevil Maskelyne, 1732—1811; Englishman, the fifth Astronomer Royal.
 ○ *crater (24 km), terraces, central peak*
SINAS (32E, 9N) Simon Sinas, 1810—76; Greek entrepreneur, patron of astronomers.   ○ *crater (12.4 km/2 260 m)*
TRANQUILLITATIS, MARE Sea of Tranquillity. Named by Riccioli. Its surface area is 438 000 sq km and therefore it can be compared with the Black Sea on the Earth. The Sea of Tranquillity contains numerous wrinkle ridges and domes, especially in its western part.
Small craters: Maskelyne B (9.2 km/1 910 m)
 Sinas E (9.2 km/1 700 m)
 Sinas A (5.8 km/1 140 m)
 Jansen Y (3.6 km/690 m)

CAUCHY

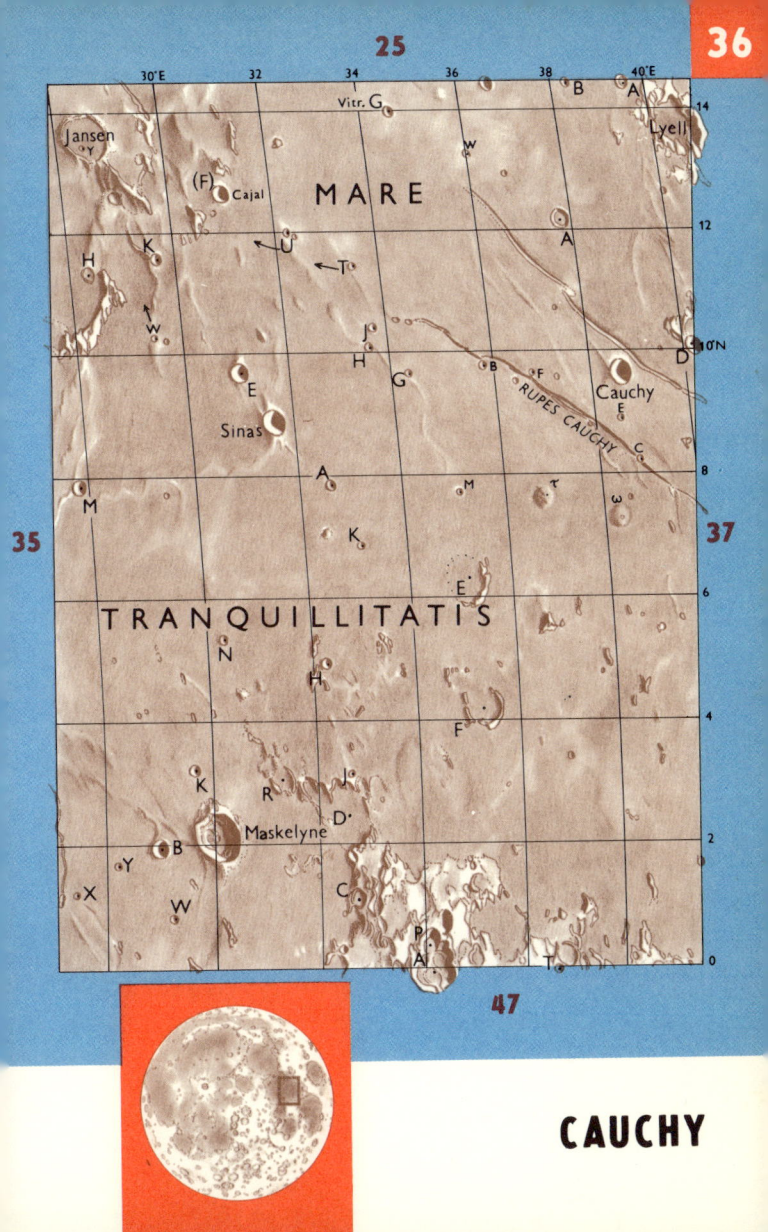

Map section 36 — Mare Tranquillitatis region

Features labeled on map:
- 25 (top), 36 (top right)
- 30°E, 32, 34, 36, 38, 40°E
- Vitr. G
- Jansen Y
- (F), Cajal
- Lyell
- MARE
- K, H, W, U, T, W, A
- Sinas, E
- J, H, G, B, F
- Cauchy, D, E, C
- RUPES CAUCHY
- 10°N
- M, A, M, τ, ω
- K, E
- 8
- 35 (left), 37 (right)
- 6
- TRANQUILLITATIS
- N, H
- F
- 4
- K, R, J
- Maskelyne, D
- Y, B
- X, W
- C
- 2
- P, A, T
- 47

CAUCHY

37 TARUNTIUS

A narrow strip of land separates Mare Tranquillitatis from Mare Fecunditatis. In the north-east there is the dark area of Mare Crisium, bordered by a mountain massif and numerous craters. Close to the terminator the edge of Mare Crisium offers a beautiful view. This area is intersected from the west by Rima Cauchy I.

CRISIUM, MARE Sea of Crises, see also maps 26, 27 and 38.
DA VINCI (44E, 10N) Leonardo da Vinci, 1452—1519; Florentine artist, poet, mathematician, architect etc; he was the first to explain the nature of the 'earthshine' on the Moon.
○ *inconspicuous crater with a disintegrated wall (31 km)*
FECUNDITATIS, MARE Sea of Fertility, see also maps 48, 49 and 59.
GLAISHER (49E, 13N) James Glaisher, 1809—1903; English meteorologist; undertook the first systematic meteorological balloon flights.
○ *prominent crater at the edge of Mare Crisium (26 km)*
LICK (53E, 12N) James Lick, 1796—1876; American financier and philanthropist; financed the construction of Lick Observatory in California and is buried in a vault at the Observatory.
○ *flooded crater (31 km)*
SECCHI (44E, 2N) Angelo Secchi, 1818—78; Italian astronomer; the first work in the field of stellar spectroscopy.
○ *inconspicuous crater with an open wall (24.5 km/1 910 m)*
SOMNI, PALUS Marsh of Sleep. A light area stretching into the Sea of Tranquillity and separated from the remaining continent in the vicinity of Mare Crisium by bright rays radiating from crater Proclus. This grey surface area stands out in the full moonlight.
TARUNTIUS (46E, 6N) Lucius Taruntius Firmanus, about 86 BC; Roman mathematician, philosopher and astrologer.
○ *crater 56 km in diameter with an outlined concentric wall and a central peak*
TRANQUILLITATIS, MARE Sea of Tranquillity, see map 36.
Small craters: Proclus A (14.7 km/2 440 m)
Taruntius E (11.3 km/2 110 m)
Taruntius EA (5 km/940 m)
Secchi A (5 km/1 050 m)
Cauchy W (3.9 km/650 m)
Abbot, Cameron, Lawrence, Tebbutt, Watts, see pp. 241—243.

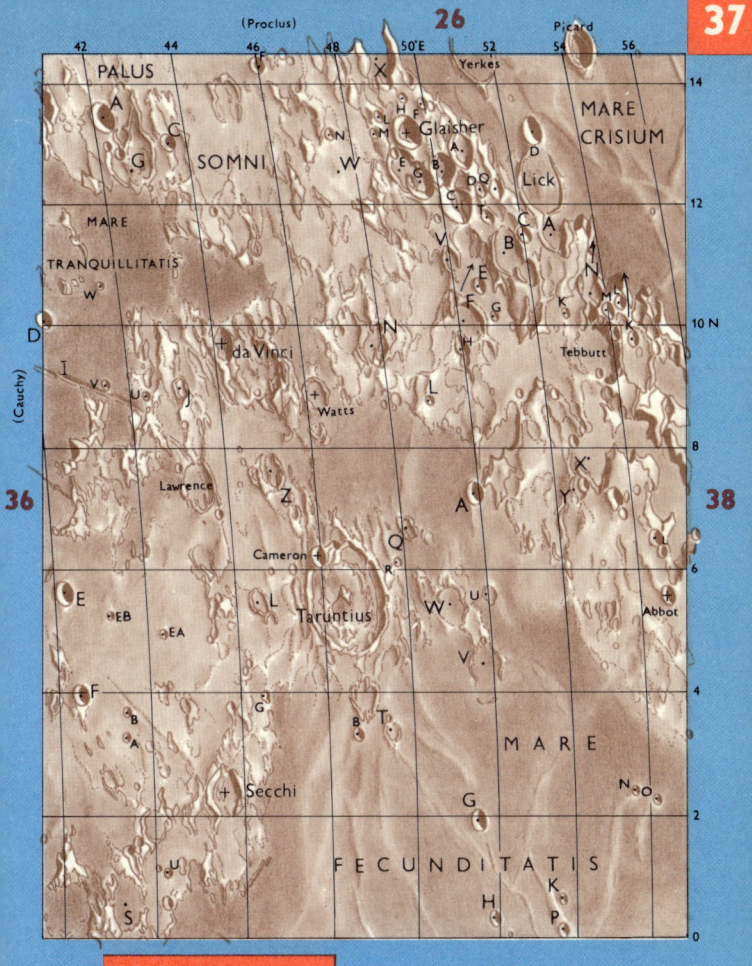

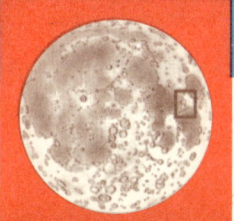

TARUNTIUS

37

26

(Proclus) Picard
42 44 46 48 50°E 52 54 56
PALUS Yerkes
A SOMNI MARE CRISIUM
G Glaisher
MARE Lick
TRANQUILLITATIS
12
da Vinci
10 N
(Cauchy) Tebbutt
Watts
8
Lawrence
36 38
Cameron
Taruntius
6
Abbot
Secchi
4
MARE
FECUNDITATIS
2

48

38 NEPER

The eastern limb of the Moon with a group of small lunar maria: Marginis, Smythii, Spumans, Undarum. This region appeared in its true shape for the first time on the historical photograph taken by Luna 3.

AGARUM, PROM. Cape Agarum. One of the few named by Hevelius.

APOLLONIUS (61E, 4N) Second half of the 3rd century BC; one of the greatest mathematicians and geometers of antiquity.
○ *flooded crater with a dark floor (52 km)*

AUZOUT (64E, 10N) Adrien Auzout, 1622–91; French astronomer; inventor of micrometer.
○ *crater (33 km) with peaks rising from its floor*

BANACHIEWICZ (80E, 5N) Tadeusz Banachiewicz, 1882–1954; Polish astronomer, selenodesist, mathematician.
○ *walled plain (92 km)*

CONDORCET (70E, 12N) Jean de Condorcet, 1743–94; French philosopher and mathematician. ○ *flooded crater (74 km)*

CRISIUM, MARE Sea of Crises, see map 27.

DUBIAGO (70E, 5N) Dmitri I. Dubiago, 1850–1918; Russian astronomer, founder of the Russian school of selenodesy.
○ *flooded crater (46 km), dark floor*

FECUNDITATIS, MARE Sea of Fertility, see map 48.

FIRMICUS (63E, 7N) Firmicus Maternus, about 330 AD; astrologer, Sicilian by origin. ○ *flooded crater (56 km)*

HANSEN (73E, 14N) Peter Andreas Hansen, 1795–1874; Danish astronomer. ○ *crater (40 km) with a central peak*

JANSKY (89E, 9N) Karl Jansky, 1905–50; American radiophysicist. ○ *crater (72 km)*

MARGINIS, MARE The 'Border Sea'.

NEPER (84E, 9N) John Napier, 1550–1617; English mathematician; inventor of logarithms (1614).
○ *very prominent crater (142 km), central massif, dark floor*

SCHUBERT (81E, 3N) Theodor F. von Schubert, 1789–1865; Russian general, directed numerous astro-geodesic activities.
○ *crater (54 km)*

SMYTHII, MARE Smyth's Sea, see also map 49. William Henry Smyth, 1788–1865; British admiral, scientific writer.
○ *circular mare, distorted along the edges*

SPUMANS, MARE The 'Foaming Sea'.

UNDARUM, MARE The Sea of Waves. Both these miniature maria were named by Franz.

Daly, Peek, Shapley, Tacchini, see pp. 242, 243.

NEPER

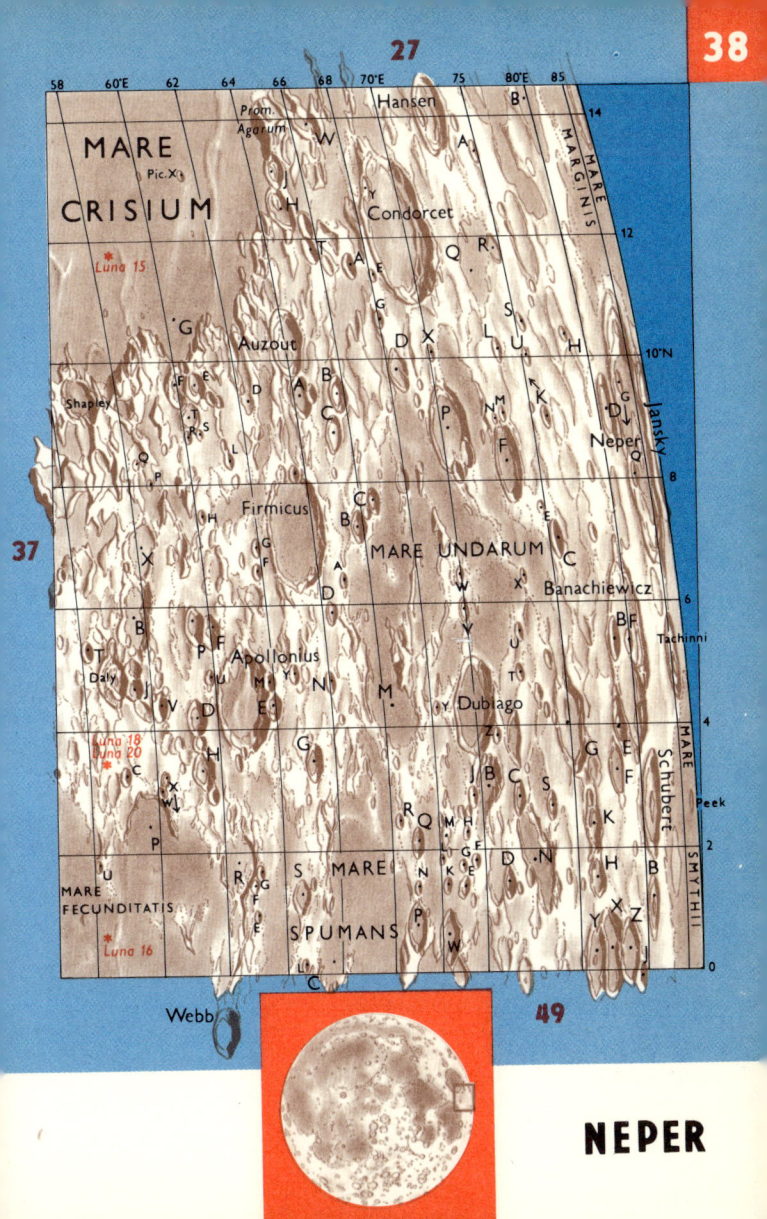

39 GRIMALDI

The western limb of the Moon, the south-western edge of Oceanus Procellarum. The north-eastern edge of the basin is formed by Mare Orientalis. The great walled plain Grimaldi, and numerous clefts, of which Rima Sirsalis is clearly visible, occur in this area.

AESTATIS, LACUS Summer Lake. Franz's original name for this area was Mare Aestatis. This lake is formed by two elongated dark patches, which stretch north of crater Crüger (see also map 50).

AUTUMNAE, LACUS Autumn Lake. Originally called Mare A. This lake is formed by dark patches situated on the inside of the Cordilleras.

CORDILLERA, MONTES Cordillera Mountains. Mountainous ring formation 900 km in diameter, which forms the inner wall of the basin Mare Orientale. Continues on map 50.

DAMOISEAU (61W, 5S) Marie Charles T. de Damoiseau, 1768—1846; French astronomer. ○ *crater (36 km)*

GRIMALDI (68 W, 5 S) Francesco Maria Grimaldi, 1618—63; Italian physicist, astronomer; prepared the map of the Moon which was used by Riccioli as a basis for nomenclature.
○ *basin with a flooded centre, which is surrounded by an inner wall 222 km in diameter; the external wall is partly disintegrated and has a diameter of 430 km.*

HARTWIG (80W, 6S) Carl E. Hartwig, 1851—1923; German astronomer. ○ *crater (80 km)*

HERMANN (57W, 1S) Jacob Hermann, 1678—1733; Swiss mathematician. ○ *ring crater (15.5 km)*

LOHRMANN (67W, 0.5S) Wilhelm G. Lohrmann, 1796—1840; German geodesist and selenographer. ○ *crater (34 km)*

PROCELLARUM, OCEANUS Ocean of Storms, see map 29.

RICCIOLI (74W, 3S) Giovanni Battista Riccioli, 1598—1671; Italian philosopher, theologian and astronomer; author of 'Almagestum Novum' in which he introduced the system of lunar nomenclature still used at the present time.
○ *walled plain (152 km)*

ROCCA (73W, 13S) Giovanni A. Rocca, 1607—56; Italian mathematician. ○ *disintegrated crater (80 km)*

SCHLÜTER (83W, 6S) Heinrich Schlüter, 1815—44; German astronomer; assistant to Bessel.
○ *prominent crater with terrace-like walls (91 km)*

SIRSALIS (60W, 12S) Gerolamo Sirsalis (Sersale), 1584—1654; Italian Jesuit, selenographer. ○ *crater (44 km)*

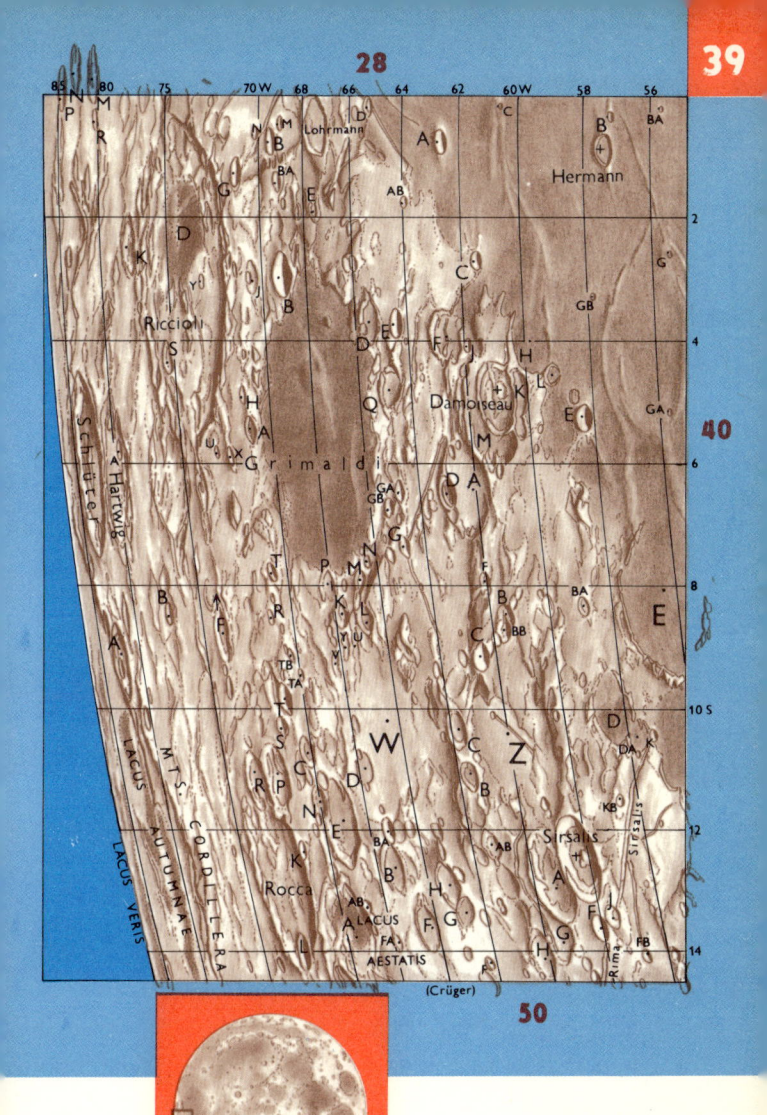

GRIMALDI

40 FLAMSTEED

The southern part of Oceanus Procellarum. The two craters Billy and Hansteen are useful points of orientation; they are separated by the bright massif Hansteen Alfa. Surveyor 1 landed close to the remains of the wall belonging to the flooded crater Flamsteed P.

BILLY (50W, 14S) Jacques de Billy, 1602—79; French Jesuit, mathematician, astronomer; rejected astrology and superstitious ideas about comets.
○ *flooded crater with a very dark floor (46 km)*

FLAMSTEED (44 W, 4 S) John Flamsteed, 1646—1720; English, the first Astronomer Royal, Director of the Greenwich Observatory; compiled the first stellar catalogue after Tycho; Flamsteed's system of numbering of the stars is still in use today. ○ *crater (21 km/2 160 m)*

HANSTEEN (52W, 12S) Christopher Hansteen, 1784—1873; Norwegian geophysicist; discovered the position of the north geomagnetic pole. ○ *crater (45 km) with hills on its floor*

LETRONNE (42W, 11S) Jean Antoine Letronne, 1787—1848; French archaeologist; an authority on ancient Egyptian civilisations.

○ *remains of a flooded walled plain 119 km in diameter. It resembles semi-circular bay in the Ocean of Storms.*

PROCELLARUM, OCEANUS Ocean of Storms, see map 29. This region contains numerous wrinkle ridges, remains of walls which belong to flooded craters, and many small hills. The neighbouring area on map 41 has a similar character.

Small craters: Flamsteed F (5.4 km/1 050 m)
Letronne B (5.2 km/1 000 m)
Flamsteed FA (3.8 km/750 m)
Letronne T (3 km/620 m)

OCEANUS

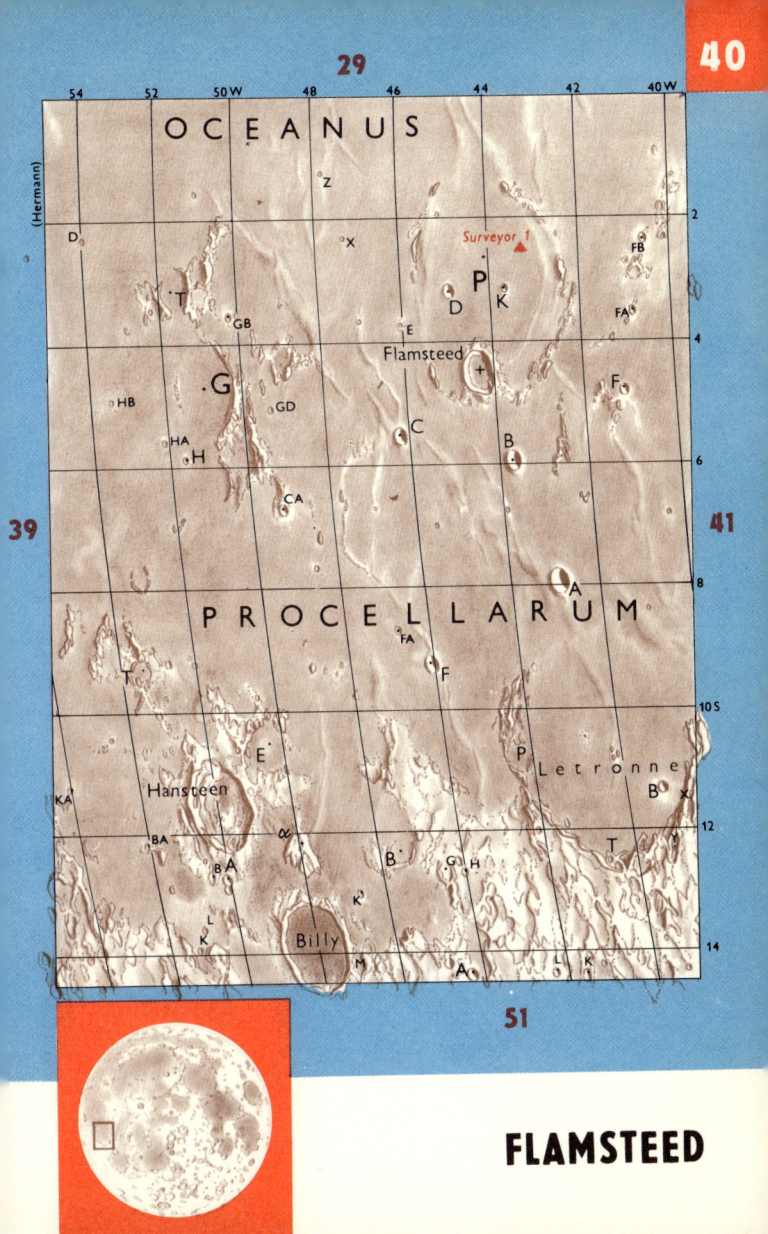

PROCELLARUM

Surveyor 1

Flamsteed

Hansteen

Letronne

Billy

FLAMSTEED

41 EUCLIDES

The south-eastern part of Oceanus Procellarum. The mountain range Riphaeus stretches along the eastern edge of the map. The massif Riphaeus Zeta is separated from the crater Lansberg by a large dome. This area contains numerous wrinkle ridges and small isolated hills also and, therefore, it is interesting for telescopic observation, especially in oblique illumination. Powerful telescopes will reveal the narrow, sinuous rille Rima Herigonius I.

EUCLIDES (30W, 7S) Euclid, about 300 BC; Greek mathematician, founder of Alexandrian mathematical school; author of the 'Elements'. ○ *very prominent and bright crater (13 km)*

HERIGONIUS (34W, 13S) Pierre Herigone, about 1644; French mathematician; his six-volume 'Cursus Mathematicus' deals also with spherical astronomy and the theory of the motion of the planets. ○ *crater (15 km/2 100 m)*

PROCELLARUM, OCEANUS Ocean of Storms, see map 29. The area north of crater Euclides F (5.2 km/1 090 m) and the neighbourhood of Herigonius contains the richest systems of wrinkle ridges on the Moon.

RIPHAEUS, MONTES Riphaean Mountains. According to ancient Greek geographers this was the mountain range from which north winds used to blow (see also map 42). The massif Riphaeus Zeta is called the Ural Mountains on the older maps of the Moon.

WICHMANN (38W, 8S) Moritz L. G. Wichmann, 1821—59; German astronomer; determined the inclination of the lunar equator and was one of the first to confirm the existence of the Moon's physical libration. ○ *ring crater (10.6 km)*

Small craters: Lansberg C (19.8 km/810 m)
Euclides B (10.3 km/2 000 m)
Lansberg B (9.9 km/2 030 m)
Lansberg G (9.9 km/270 m)
Wichmann C (2.8 km/490 m)

EUCLIDES

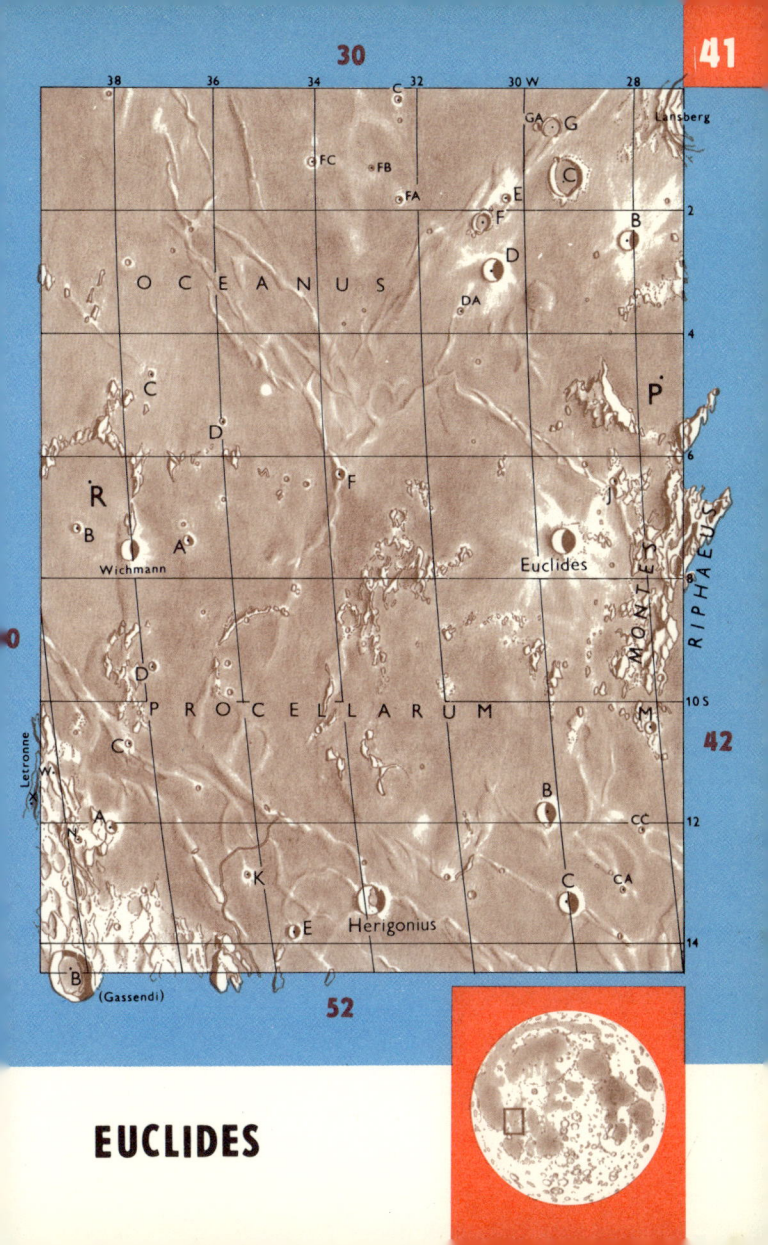

42 FRA MAURO

The south-eastern edge of the Ocean of Storms stretches to this section of the Moon and separates craters Lansberg and Fra Mauro. The bottom half of the map is occupied by Mare Cognitum. Although this area is seemingly uninteresting, it contains a number of topographically and geologically interesting features, and several landings on the Moon have been made in this region. This is the area where crash landings of the probes Ranger 7 and Luna 5 took place; Surveyor 3 and two Apollo expeditions landed here, Apollo 12 close to Surveyor 3 and Apollo 14 in the hills at the edge of the crater Fra Mauro.

BONPLAND (17W, 8S) Aimé Bonpland, 1773—1858; French botanist who accompanied Humboldt on his expeditions to Mexico and Columbia.
  ○ *remains of walled plain 60 km in diameter with narrow clefts on the floor*
COGNITUM, MARE Known Sea. Named in 1964 after the successful flight of Ranger 7, which transmitted to the Earth the first detailed television photographs of the surface of a lunar mare. The seemingly smooth and flat surface of the mare is rippled by numerous craters. The photographs taken by Ranger 7 disclosed some pits which are 10^{-3} of the size of markings which are visible through large telescopes. A new stage in the exploration of the Moon was started here.
DARNEY (24W, 15S) Maurice Darney, 1882—1958; Frenchman, observer of the Moon.   ○ *crater (15 km/2 620 m)*
FRA MAURO (17W, 6S) Died in 1459; Venetian geographer; prepared a map of the World (1457).
  ○ *remains of walled plain (94 km). Its centre is intersected by clefts which run in the meridian direction.*
LANSBERG (27W, 0.3S) Philippe van Lansberg, 1561—1632; Belgian physician and astronomer, author of a treatise on the methods of using astrolabes and gnomons.
                 ○ *prominent crater (40 km/3 110 m)*
PROCELLARUM, OCEANUS Ocean of Storms, see map 29. Bright rays from the crater Copernicus extend into this area.
RIPHAEUS, MONTES, see map 41.
  Small craters: Darney C (13.3 km/2 330 m)
                 Fra Mauro B (7 km/1 380 m)
                 Bonpland E (6.8 km/1 330 m)

FRA MAURO

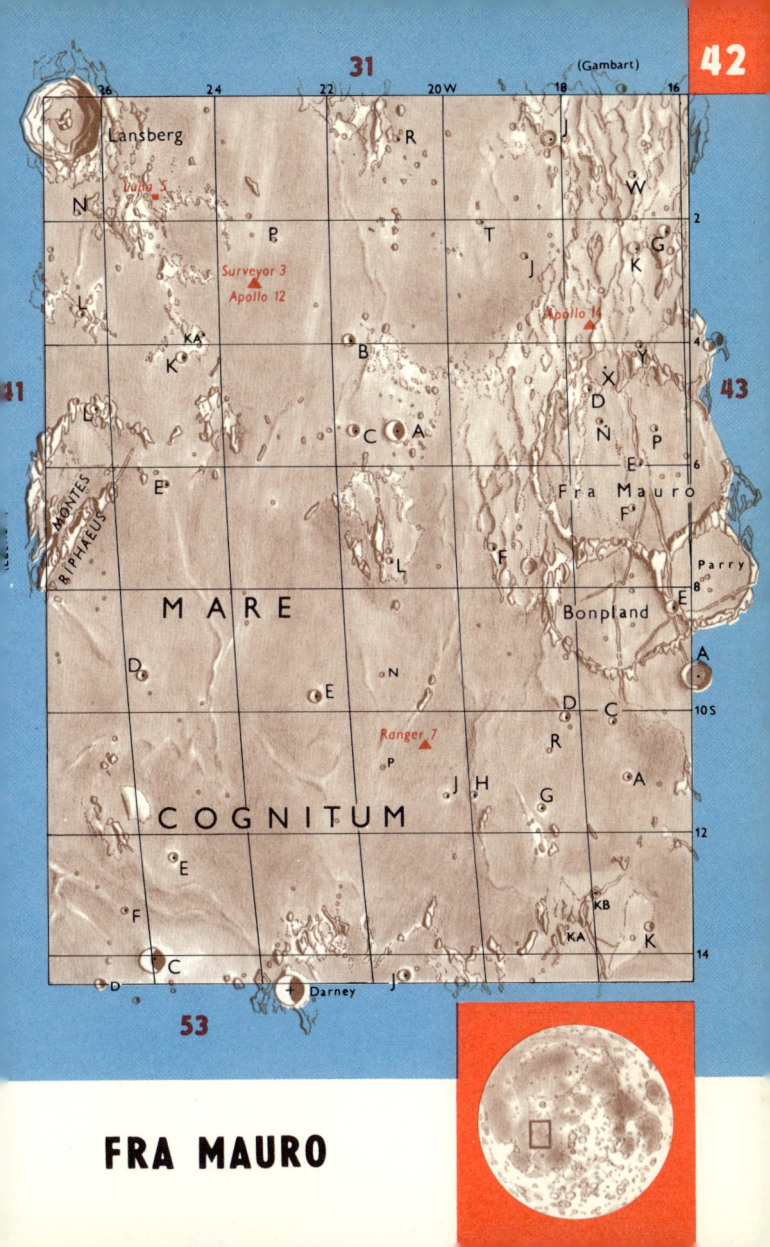

43 LALANDE

Predominantly mare area, which is penetrated by Mare Nubium from the south and in the west adjoins the three craters Bonpland, Parry and Fra Mauro. The edge of a vast continent, which shows traces of the development of the basin of Mare Imbrium (grooves, valleys, clefts), protrudes to this region from the east. Crater Davy Y contains an interesting chain of crater pits.

DAVY (8W, 12S) Sir Humphry Davy, 1778—1829; English chemist; inventor of the miners' safety-lamp. ○ *crater (35 km) with its wall intersected by crater Davy A (15 km)*

GUERICKE (14W, 12S) Otto von Guericke, 1602—86; German natural philosopher; in 1654 demonstrated the existence of atmospheric pressure by the well-known experiment with the 'Magdeburg's hemispheres'.
○ *remains of a walled plain (58 km)*

LALANDE (9W, 4S) Joseph J. le Francois de Lalande, 1732—1807; French astronomer; Director of the Paris Observatory. ○ *prominent crater (24 km/2 590 m), terraces, centre of bright rays.*

MÖSTING (6W, 1S) Johann S. von Mösting, 1759—1843; Danish statesman; helped to found the periodical 'Astronomische Nachrichten'.
○ *crater with terrace-like walls (26 km/2 760 m)*

MÖSTING A (5°09'50" W, 3°10'47" S) Bright, small ring crater (13 km/2 700 m); the zero point in the network of selenodesic coordinates.

PALISA (7W, 9S) Johann Palisa, 1848—1925; Austrian astronomer; discovered 127 small planets.
○ *crater (33 km), disintegrated wall, open to south-west*

PARRY (16W, 8S) Sir William E. Parry, 1790—1855; English admiral and arctic explorer.
○ *crater with a flooded floor (46 km/560 m)*

TURNER (13W, 1S) Herbert Hall Turner, 1861—1930; English astronomer; participated in formulation of the international lunar nomenclature. ○ *crater (11.8 km/2 630 m)*

Small craters:
Lalande A (13.2 km/2 600 m)   Davy C (3.4 km/540 m)
Guericke D (7.6 km/1 500 m)   Parry D (2.8 km/330 m)
Fra Mauro R (3.4 km/650 m) on the top of the hill Fra Mauro Eta

LALANDE

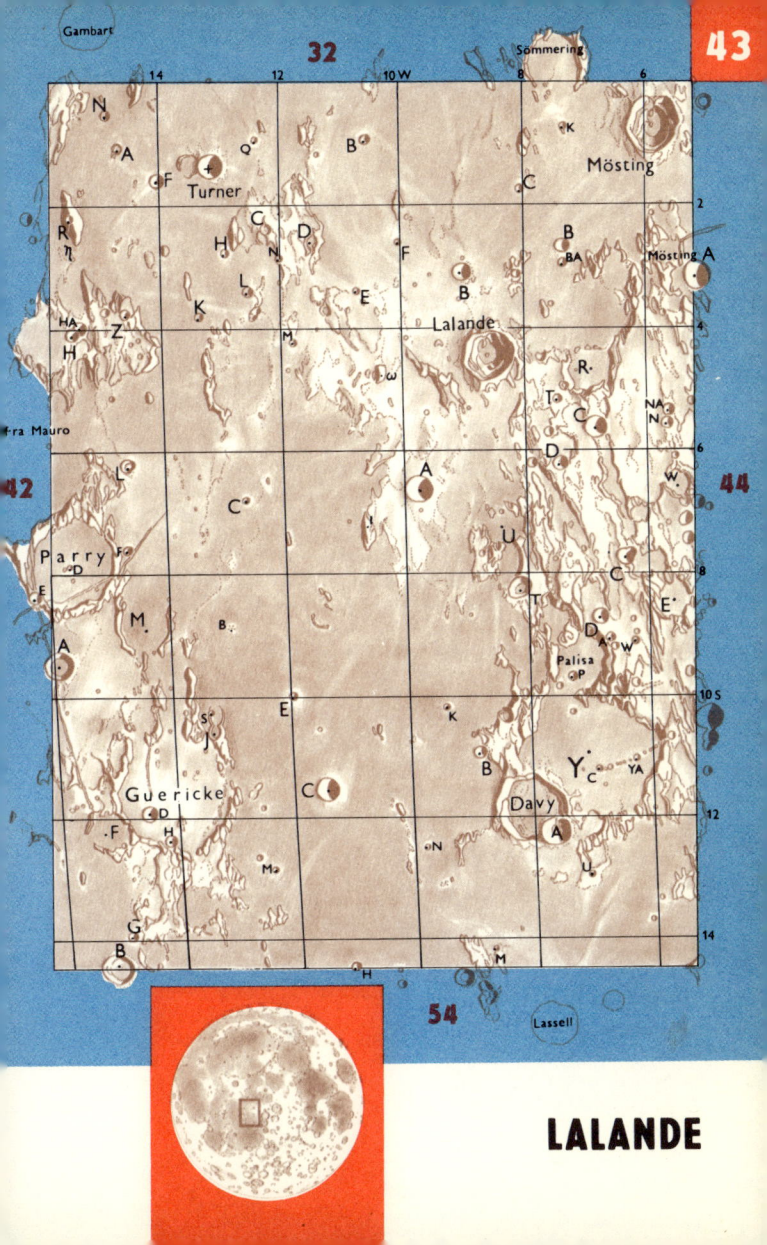

44 PTOLEMAEUS

The central part of the near side of the Moon with large walled plains. The dark surface of Sinus Medii, bordered with clefts, stretches to this area from the north. This region, similarly to the area shown on map 43, displays a system of valleys and clefts directed towards Mare Imbrium.

ALBATEGNIUS (4E, 11S) Muhammed ben Geber al Batani, about 850—929; Arabian prince and astronomer.
○ *crater (136 km)*

ALPHONSUS (3W, 14S) Alfonso X, 'El Sabio', 1223—84; King of Castile, responsible for production of the 'Alphonsine Tables'. ○ *crater (118 km/2 730 m); its floor has clefts, a central peak, crater pits with dark halo*

FLAMMARION (4W, 3S) Nicolas Camille Flammarion, 1842—1925; French astronomer; popularizer of astronomy.
○ *walled plain (75 km); Mösting A rises from its western wall*

GYLDÉN (0.3E, 5S) Johan August Hugo Gyldén, 1841—96; Swedish astronomer. ○ *disintegrated crater (47 km)*

HERSCHEL (2W, 6S) Sir William Herschel, 1738—1822; English astronomer, German by birth; discovered Uranus; pioneered stellar astronomy; discovered 2 500 nebulae and galaxies. ○ *prominent crater, terraces (41 km/3 770 m)*

HIPPARCHUS (5E, 6S) About 140 BC; Greek astronomer; prepared the first stellar catalogue.
○ *considerably disintegrated walled plain (150 km/3 320 m)*

KLEIN (3E, 12S) Hermann J. Klein, 1844—1914; German selenographer. ○ *crater (44 km/1 460 m)*

MEDII, SINUS Central Bay, see map 33.

MÜLLER (2E, 8S) Karl Müller, 1866—1942; Austrian selenographer. ○ *elongated crater (24 × 20 km)*

OPPOLZER (0.5W, 2S) Theodor Egon von Oppolzer, 1841—86; Austrian astronomer. ○ *remains of a crater (43 km), cleft*

PTOLEMAEUS (2W, 9S) Claudius Ptolemaeus, about 130 AD; Greek astronomer, author of the 'Almagest', geocentric model of the Universe. ○ *very prominent ringed plain (153 km/ 2 400 m) with numerous pits and depressions on its floor*

RÉAUMUR (1E, 2S) René Antoine Ferchault de Réaumur, 1683—1757; French naturalist; 'Réaumur's thermometer'.
○ *remains of a crater (53 km)*

SEELIGER (3E, 2S) Hugo von Seeliger, 1849—1924; German astronomer. ○ *crater (8.5 km/1 800 m)*

SPÖRER (2W, 4S) Friedrich W. G. Spörer, 1822—95; German astronomer. ○ *indistinct, shallow crater (28 km/310 m)*

PTOLEMAEUS

45 ANDĚL

A continental region with a number of considerably disintegrated craters. Tectonics of the basin of Mare Imbrium also belong to this area.

ABULFEDA (14E, 14S) Ismail Abu'l-Fida, 1273—1331; Syrian prince, geographer, astronomer. ○ *crater (62 km/3 110 m)*
ANDĚL (12E, 10S) Karel Anděl, 1884—1947; Czech teacher and selenographer; 'Mappa Selenographica' (1926).
○ *polygonal crater (34 km), wall is open in the south*
BURNHAM (7E, 14S) Sherburne Wesley Burnham, 1838—1921; American amateur astronomer; discovered over 1200 double stars. ○ *inconspicuous crater with a disintegrated wall (22 km)*
DESCARTES (16E, 12S) René Descartes, 1596—1650; French philosopher and mathematician. ○ *crater (48 km)*
DOLLOND (14E, 10S) John Dollond, 1706—61; English optician. ○ *crater (11.1 km/1 580 m)*
HALLEY (6E, 8S) Edmond Halley, 1656—1742; English astronomer; on the basis of Newton's gravitational law proved the periodic character of a comet's orbit, which was named after him. ○ *crater (36 km/2 510 m)*
HIND (7E, 8S) John Russel Hind, 1823—95; English astronomer. ○ *crater (29 km/2 980 m); in line with the craters Hipparchus C (17 km/2 940 m) and L (13 km/2 630 m)*
HIPPARCHUS, see map 44.
HORROCKS (6E, 4S) Jeremiah Horrocks, 1619—41; English astronomer; first to observe a transit of Venus (1639).
○ *prominent crater (30 km/2 980 m)*
LADE (10E, 1S) Heinrich E. von Lade, 1817—1904; German banker and amateur astronomer. ○ *remains of a flooded crater (56 km). The crater Lade B is filled to the brim.*
PICKERING (7E, 3S) Edward Charles Pickering, 1846—1919; American astronomer. ○ *crater (16 km/2 740 m)*
RITCHEY (8E, 11S) George Willis Ritchey, 1864—1919; American astronomer. ○ *crater (24 km/1 300 m)*
SAUNDER (9E, 4S) Samuel A. Saunder, 1852—1912; English selenographer; catalogue of positions of 3 000 points on the Moon. ○ *crater with a low, irregular wall (45 km)*
THEON JUNIOR (16E, 2S) Theon of Alexandria, about 380 AD; last astronomer of the Alexandrian school, father of Hypatia (see plate 46). ○*prominent crater (18.6 km/3 580 m)*
THEON SENIOR (15E, 1S) Theon of Smyrna, about 100 AD; Greek mathematician and astronomer.
○ *crater (18.2 km/3 470 m)*

ANDĚL

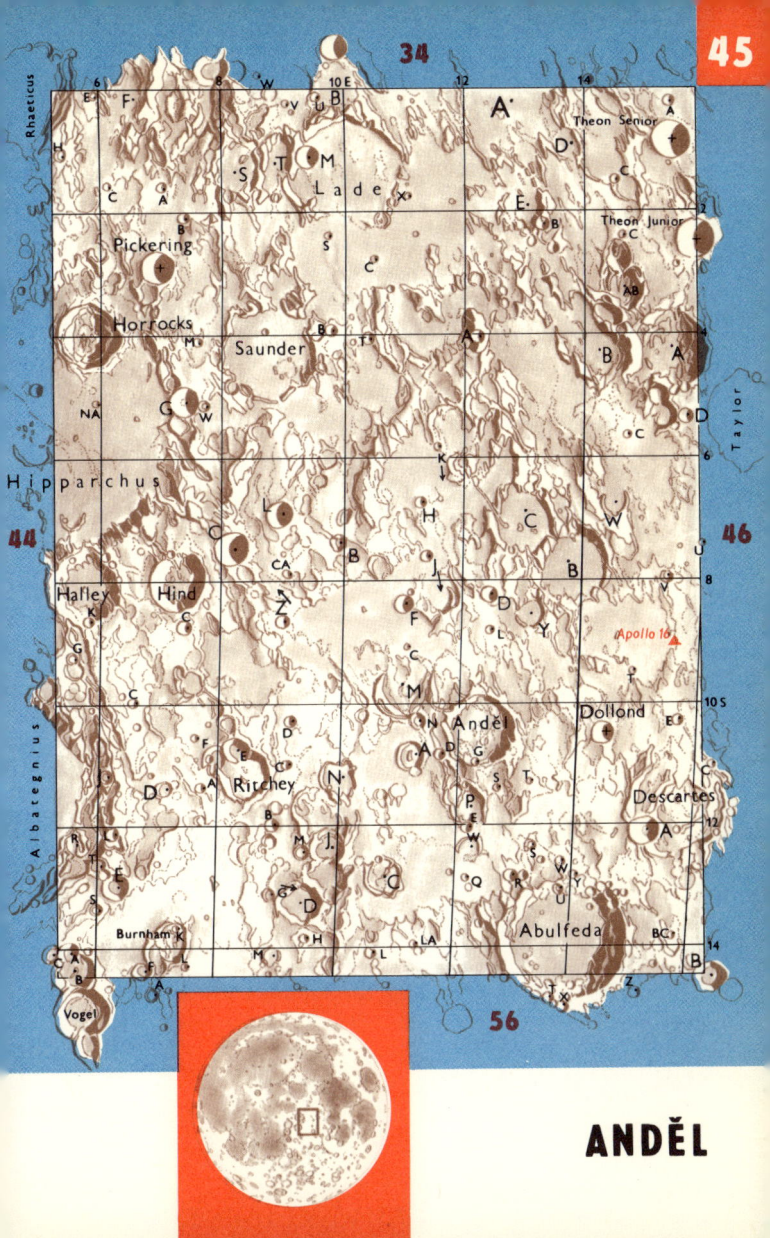

46 THEOPHILUS

From the north, this area is penetrated by the edge of Mare Tranquillitatis with the shallow rille Rima Hypatia I. The large crater Theophilus is one of the most attractive formations on the Moon. It forms a characteristic trio with the craters Cyrillus and Catharina (map 57). Craters Kant B and Zöllner E are situated on the opposite edges of an extensive plateau which resembles a filled crater. The mountainous massif situated to the north-east of Kant is 4 000 m high.

ALFRAGANUS (19E, 5S) Muhammed ibn Ketir al Fargani, about 840; Arabian astronomer.
○ *irregular crater (21 km/3 830 m)*
CYRILLUS (24E, 13S) St. Cyril, died in 444 AD; Bishop of Alexandria after Theophilus.
○ *ring mountain chain with a disintegrated wall (93 km)*
DELAMBRE (18E, 2S) Jean Baptiste Joseph Delambre; 1749—1822; French astronomer, theoretician; measured arc of meridian as basis for fixing length of the metre.
○ *prominent crater, terraces (53 km)*
HYPATIA (23E, 4S) Died in 415 AD; daughter of Theon of Alexandria; astronomer and mathematician.
○ *irregular crater (41 × 28 km)*
KANT (20E, 11S) Immanuel Kant, 1724—1804; German philosopher, propounded nebular hypothesis of the evolution of the solar system.
○ *regular crater (32 km/3 120 m) with a central peak*
MOLTKE (24E, 1S) Helmuth Carl Bernhard, Graf von Moltke, 1800—91; Prussian Field Marshall and statesman; secured publication of Schmidt's map of the Moon.
○ *crater with a bright rim (6.5 km/1 310 m)*
TAYLOR (17E, 5S) Brook Taylor, 1685—1731; English mathematician and philosopher. ○ *elliptic crater (41 × 34 km)*
THEON JUNIOR, see map 45.
THEOPHILUS (26E, 11S) St. Theophilus, died in 412 AD; Alexandrian bishop from 385 AD.
○ *crater (100 km/4 400 m); the ridge of the wall rises 1 200 m above the neighbouring landscape, its central mountain chain is 2 000 m high.*
ZÖLLNER (19E, 8S) Johann Karl Friedrich Zöllner, 1834—82; German astronomer, invented astro-photometer.
○ *elongated crater (46 × 36 km) with a disintegrated wall*

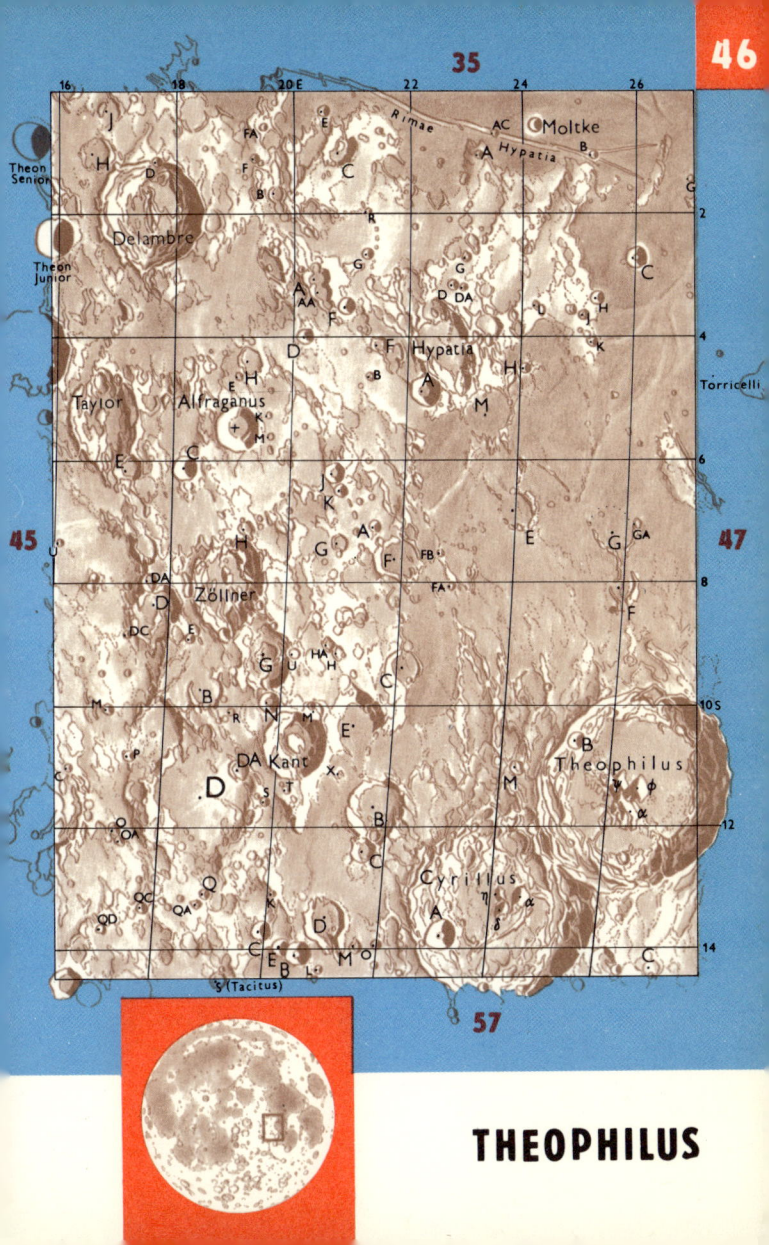

THEOPHILUS

47 CAPELLA

This continental promontory separates Mare Tranquillitatis from Mare Nectaris. The continent is traversed by the rilles Rimae Gutenberg. On the edge of M. Nectaris there is a prominent pair of craters, Capella and Isidorus. The double pear-shaped crater Torricelli and the small crater Censorinus, which is one of the brightest objects on the Moon, also contribute to the local peculiarities.

CAPELLA (35E, 8S) Marianus Capella, 5th century AD; Carthaginian lawyer; Copernicus refers to his theory that Mercury and Venus are in orbit around the Sun, and that the Sun with the remaining planets revolve around the Earth.
○ *crater (45 km)*

CENSORINUS (32.7E, 0.4S) About 238 AD; in his letter 'De Die Natali' deals with the influences of stars on chronology.
○ *small crater (3.8 km), exceptionally bright*

DAGUERRE (34E, 12S) Louis Daguerre, 1789—1851; French painter; inventor of a photographic system, now known as 'daguerrotype'.
○ *inexpressive ring depression 45 km in diameter*

GAUDIBERT (38E, 11S) Casimir M. Gaudibert, 1823—1901; French amateur astronomer, selenographer.
○ *inconspicuous crater (28 km), divided by central peaks and mountain ranges*

ISIDORUS (34E, 8S) St. Isidor of Seville, about 570—636 AD; Bishop of Seville; marginally interested in astronomy.
○ *crater (39 km/1 580 m)*

MÄDLER (30E, 11S) Johann Heinrich Mädler, 1794—1874; German selenographer, author of monograph 'Der Mond' and of a map of the Moon. ○ *regular crater (28 km/2 670 m)*

NECTARIS, MARE Sea of Nectar, see also map 58.

TORRICELLI (28E, 5S) Evangelista Torricelli, 1608—47; Italian natural philosopher, contemporary of Galileo, inventor of the mercury barometer.
○ *crater, 20 km in diameter; the wall in the west is open and linked with a smaller crater so that the whole formation is pear-shaped*

TRANQUILLITATIS, MARE Sea of Tranquillity, see maps 35 and 36.

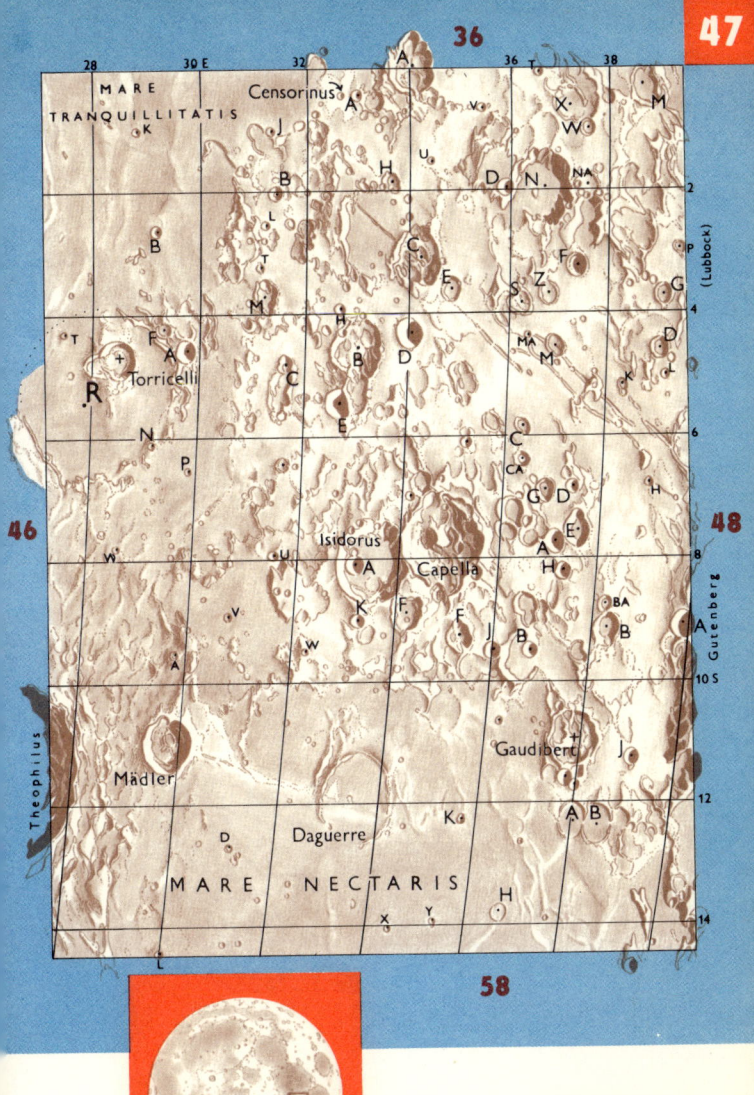

CAPELLA

MARE TRANQUILLITATIS
Censorinus
Torricelli
Isidorus
Capella
Mädler
Gaudibert
Daguerre
MARE NECTARIS
Theophilus
(Lubbock)
Gutenberg

48 MESSIER

In the vast area of Mare Fecunditatis, which is furrowed by wrinkle ridges, there is the well-known pair of craters Messier and Messier A, which is the centre of two bright rays radiating to the west. In oblique illumination 'ghost craters' can be seen close to Goclenius A and also the system of rilles along Gutenberg.

BELLOT (48E, 12S) Joseph R. Bellot, 1826—53; French seaman, participated in two Antarctic expeditions. ○ *crater (17 km)*
CROZIER (51E, 14S) Francis Rawdon Moira Crozier, 1796—1848; English naval captain; participated in the arctic expeditions of Parry and Rosse. ○ *flooded crater (22 km)*
FECUNDITATIS, MARE Sea of Fertility. Irregularly shaped mare with a surface of 337 000 sq km (smaller than the Caspian Sea on the Earth).
GOCLENIUS (45E, 10S) Rudolf Göckel, 1572—1621; German physician and mathematician.
  ○ *irregular crater (54 × 66 km) with clefts on its floor*
GUTENBERG (41E, 9S) Johann Gutenberg, about 1398—1468; German, invented letter-printing.
  ○ *crater 71 km in diameter; its wall in the east is broken up by flooded crater E, which in the east is connected with crater C. Its floor has a number of peaks and a cleft. In the south-western wall lies crater Gutenberg A (15 km/3 430 m).*
LUBBOCK (42E, 4S) Sir John William Lubbock, 1803—65; English mathematician and astronomer. ○ *crater (14.5 km)*
MAGELHAENS (44E, 12S) Fernao de Magalhaes (Magellan), 1480—1521; Portuguese navigator; he was first to sail round Cape Horn and his flotilla completed the first circumnavigation of the world. ○ *flooded crater (38 km), dark floor*
MESSIER (48E, 2S) Charles Messier, 1730—1817; French astronomer; discovered several comets; compiled the catalogue of star clusters and nebulae known as Messier's catalogue.
  ○ *oval crater (9 × 11 km)*
MESSIER A (previously W. H. Pickering)
  ○ *double crater (13 × 11 km), which is the centre of two bright rays radiating to the west*
PYRENAEUS, MONTES Pyrenees. Mädler's name for the mountain range situated south of crater Gutenberg; see also maps 47, 58 and 59.

MESSIER

49 LANGRENUS

The eastern edge of Mare Fecunditatis and the eastern limb of the Moon are intersected by a wide and dense crater field, in which orientation is not easy. Crater Langrenus offers beautiful sight through the telescope. During favourable librations the dark surface of Mare Smythii is visible along the limb of the Moon.

ANSGARIUS (79E, 13S) St Ansgar, 801—64; Church dignitary and missionary, born in Picardia.
   ○ *prominent crater with terrace-like walls (94 km)*
FECUNDITATIS, MARE Sea of Fertility, see map 48.
GILBERT (76E, 3S) Grove Karl Gilbert, 1843—1918; American geologist.   ○ *walled plain (107 km)*
KAPTEYN (71E, 11S) Jacobus Cornelius Kapteyn, 1851—1922; Dutch astronomer; model of Milky Way; studied the motions and parallaxes of stars.   ○ *prominent crater (49 km)*
KÄSTNER (79E, 7S) Abraham Gotthelf Kästner, 1719—1800; German mathematician and physicist.
   ○ *walled plain (119 km)*
LAMÉ, see map 60.
LANGRENUS (61E, 9S) Michel Florent van Langren, about 1600—75; Belgian engineer and mathematician; drew the first map of the Moon with the names of formations.
   ○ *very prominent crater with terrace-like walls, and hills and central peaks on its floor (132 km)*
LA PÉROUSE (77E, 11S) Jean Francois de Galoup, Comte de la Pérouse, 1741—88; French navigator; he led an expedition round the world, which became lost in the Pacific.
   ○ *crater (78 km)*
LOHSE (60E, 14S) Oswald Lohse, 1845—1915; German astronomer; photographed planets, prepared maps of Mars.
   ○ *crater (42 km) by the northern edge of crater Vendelinus*
MACLAURIN (68E, 2S) Collin Maclaurin, 1698—1746; English mathematician, continued Newton's work in the field of mathematics.   ○ *inconspicuous crater (50 km), central peak*
SMYTHII, MARE see map 38.
WEBB (60E, 1S) Thomas William Webb, 1806—85; English amateur astronomer and draftsman.
   ○ *crater (23 km); Luna 16 landed north of this area*
Haldane, Houtermans, Kiess, Kreiken, Runge, Widmannstätten, see 242, 243.

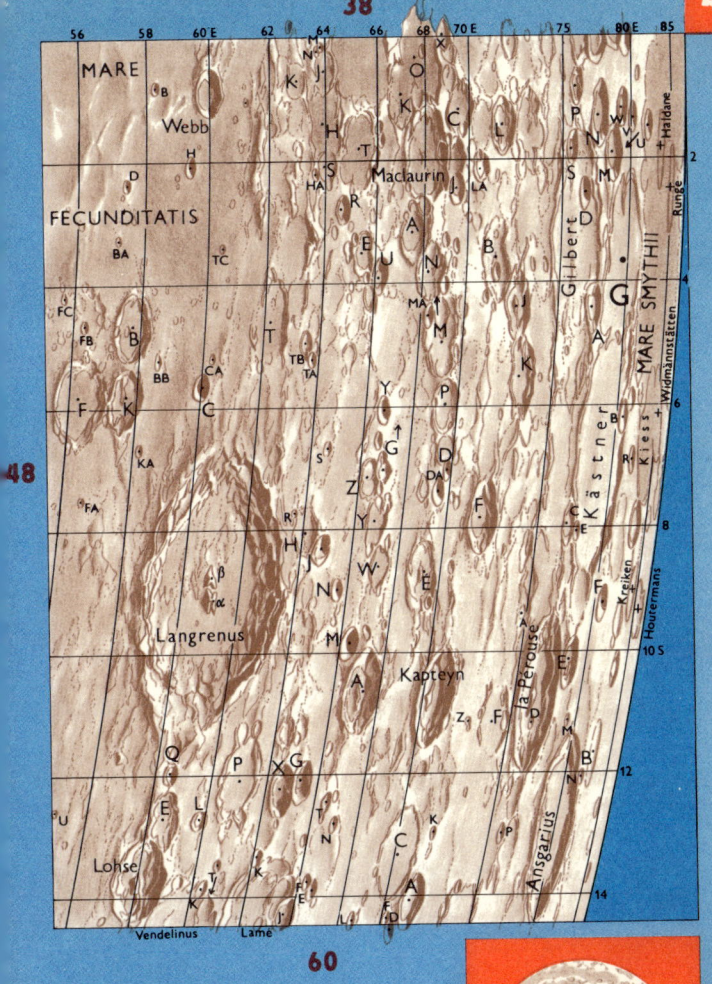

LANGRENUS

50 DARWIN

The western limb of the Moon, in which a part of the remarkable lunar basin of Mare Orientalis is situated. Interesting object for observation is the wide Rima Sirsalis, which continues on map 39.

AESTATIS, LACUS, see map 39.

AUTUMNAE, LACUS, see map 39.

BYRGIUS (65W, 25S) Joost Bürgi, 1552—1632; Swiss clockmaker, excellent mechanic (sextant Tycho Brahe).
○ *crater (90 km). Byrgius A is the centre of bright rays*

CORDILLERA, MONTES Cordillera Mountains. The eastern part of the circular mountain chain 900 km in diameter (see map 39).

CRÜGER (67W, 17S) Peter Crüger, 1580—1639; German mathematician, teacher of Hevelius.
○ *crater with a very dark floor (45 km)*

DARWIN (69W, 20S) Charles Robert Darwin, 1809—82; English naturalist, author of the theory of the evolution through natural selection. ○ *disintegrated walled plain (132 km)*

EICHSTÄDT (78W, 23S) Lorenz Eichstädt, 1596—1660; German physician, mathematician and astronomer.
○ *prominent crater on the edge of the Cordilleras (49 km)*

KRASNOV (80W, 30S) Alexander V. Krasnov, 1866—1907; Russian astronomer; measured lunar librations with heliometer.
○ *crater (44 km)*

LAMARCK (70W, 22S) Jean Baptiste Pierre Antoine de Monet Lamarck, 1744—1829; French naturalist; founder of zoology of invertebrates. ○ *considerably disintegrated crater (110 km)*

NICHOLSON (85W, 26S) Seth Barnes Nicholson, 1891—1963; American astronomer. ○ *crater in the Mt. Rook (36 km)*

ORIENTALE, MARE Eastern Sea. Flooded centre (about 250 km in diameter) of the youngest lunar basin. The whole of this mare is situated on the far side of the Moon and its edges are visible only during favourable librations.

PETTIT (86W, 27S) Edison Pettit, 1890—1962; American astronomer; carried out research on the solar prominences.
○ *crater (36 km), which forms a pair with Nicholson*

ROOK, MONTES Rook Mountains. Lawrence Rooke, 1622—66; English astronomer, observer of Jupiter's satellites.
○ *one of the inner circular mountain chains which surround the basin of Mare Orientale*

VERIS, LACUS Spring Lake. A narrow mare on the inner edge of Montes Rook.

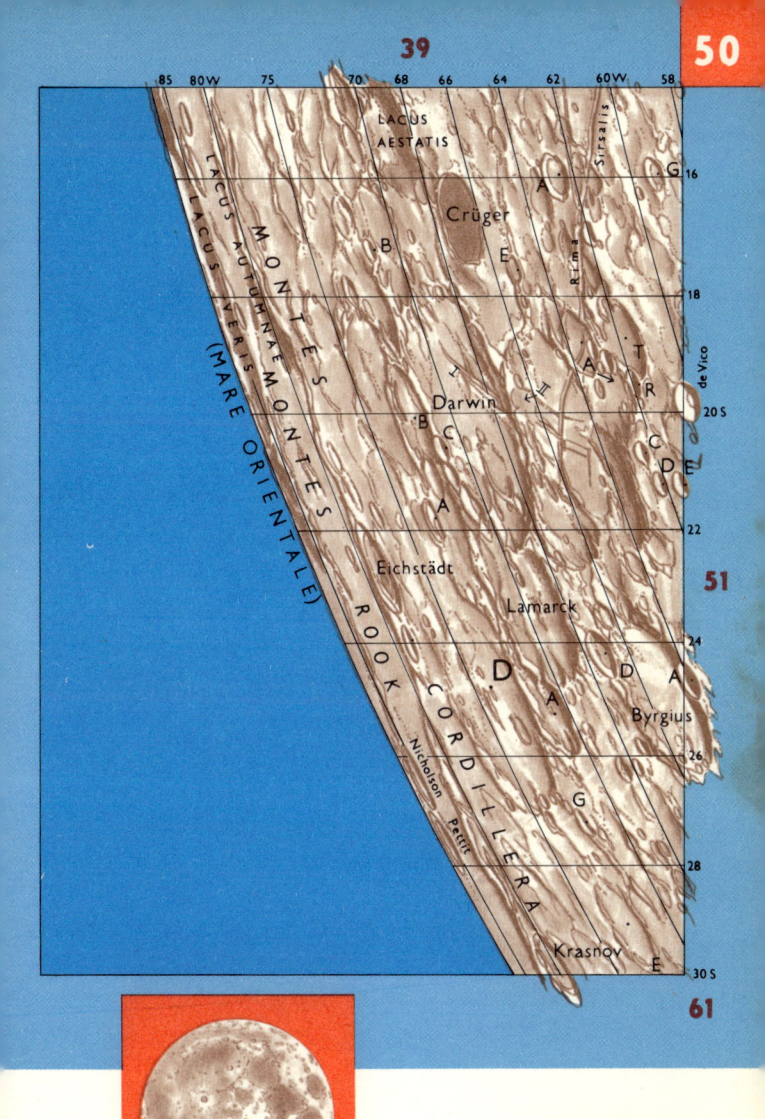

DARWIN

Map features

- LACUS AESTATIS
- Crüger
- MONTES
- LACUS AUTUMNI
- LACUS VERIS
- (MARE ORIENTALE)
- Darwin
- de Vico
- ROOK
- CORDILLERA
- Eichstädt
- Lamarck
- Byrgius
- Nicholson
- Pettit
- Krasnov

51 MERSENIUS

The hilly region between the western edge of Mare Humorum and the western limb of the Moon. Clefts and furrows run parallel with the edge of Mare Humorum.

CAVENDISH (54W, 25S) Henry Cavendish, 1731–1810; English natural philosopher; discovered hydrogen; 'Cavendish's experiment' with torsion balance to determine the density of the Earth.
○ *crater (56 km) with a wall divided by crater Cavendish E*
DE GASPARIS (51W, 26S) Annibale de Gasperis, 1819–92; Italian astronomer; discovered 6 small planets.
○ *flooded crater (32 km) with clefts in its floor*
DE VICO (60W, 20S) Francesco de Vico, 1805–48; Italian astronomer; observed Venus and discovered 6 comets.
○ *crater (19 km)*
FONTANA (57W, 16S) Francesco Fontana, about 1585–1656; Italian lawyer and amateur astronomer; observer of planets.
○ *crater (31 km)*
HENRY, FRÈRES Henry Brothers: Paul Henry (57W, 24S) 1848–1905; Prosper Henry (59W, 24S) 1849–1903. French astronomers, worked in the field of astro-photography in order to produce a photographic map of the sky (Carte du Ciel); constructed large refractors.
○ *crater pair: Paul Henry — crater (41 km) Prosper Henry — crater (42 km)*
HUMORUM, MARE Sea of Moisture, see map 52.
LIEBIG (48W, 24S) Justus, Baron von Liebig, 1803–73; German chemist; invented a new process for silvering glass, which was suitable for mirrors used in astronomical telescopes.
○ *crater (38 km)*
MERSENIUS (49W, 22S) Marian Mersenne, 1588–1648; French theologian, mathematician, natural philosopher.
○ *flooded crater with a convex floor (82 km)*
PALMIERI (48W, 29S) Luigi Palmieri, 1807–96; Italian mathematician and geophysicist. ○ *flooded crater (40 km)*
VIETA (56W, 29S) François Viète, 1540–1603; French lawyer and mathematician. ○ *crater (87 km)*
ZUPUS (52W, 17S) Giovanni B. Zupi, about 1590–1650; Italian Jesuit and astronomer.
○ *remains of a flooded crater (35 km)*

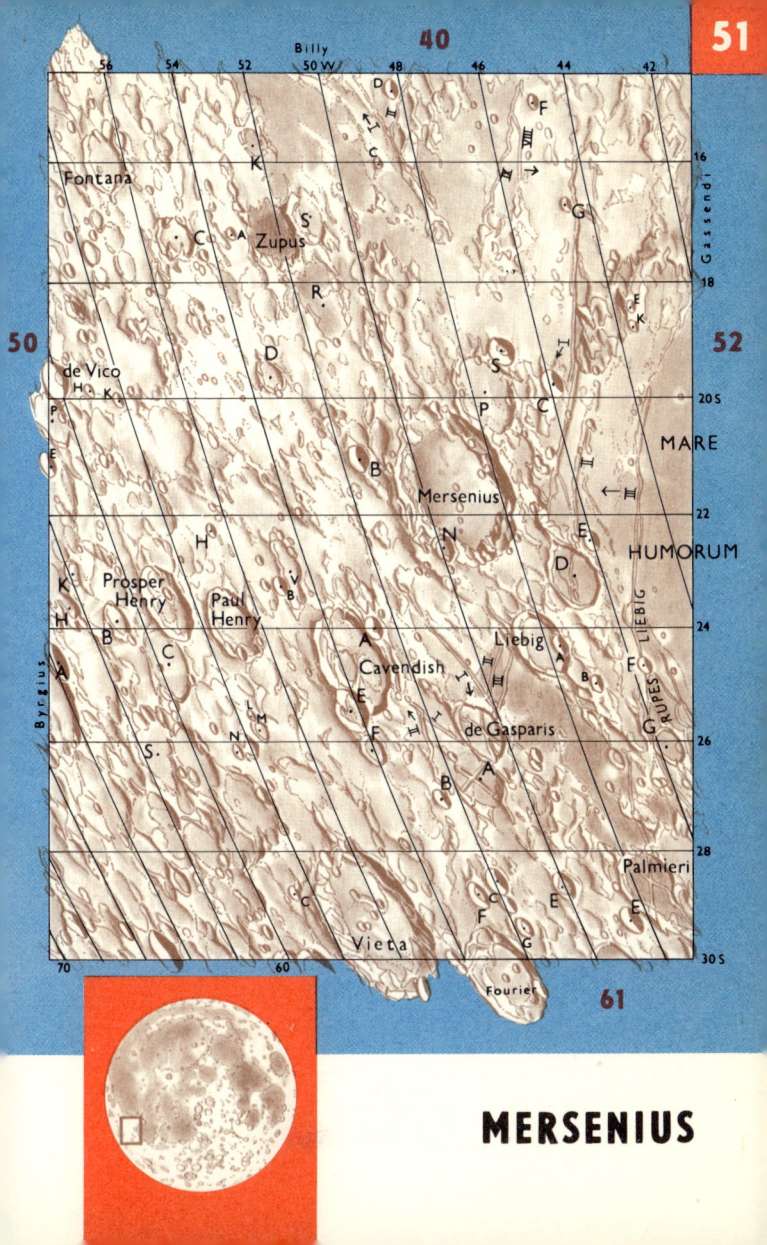

MERSENIUS

52 GASSENDI

The basin of Mare Humorum and an interesting ringed plain Gassendi, resembling a ring with a precious stone, are two of the most attractive features of the south-western quadrant on the near side of the Moon. The system of concentric wrinkle ridges in M. Humorum and the neighbouring rilles Rimae Hippalus are good objects for telescopic observation.

AGATHARCHIDES (31W, 20S) End of 2nd century BC; Greek geographer and historian. ○ *flooded crater (49 km/1 180 m)*

DOPPELMAYER (41W, 28S) Johann Gabriel Doppelmayer, 1671—1750; German mathematician and astronomer; made a map of the Moon.
○ *greatly eroded crater (64 km)*

GASSENDI (40W, 18S) Pierre Gassendi, 1592—1655; French theologian, mathematician and astronomer; supported Copernicus's theories, exchanged letters with Kepler and Galilei; first to observe the transit of Mercury across the Sun in 1631, which was forecast by Kepler.
○ *prominent ringed plain (110 km/1860 m) with numerous clefts hills and central mountains on its floor. The wall is intersected by the crater Gassendi A (33 km/3 600 m).*

HIPPALUS (30W, 25S) About 120 AD; Greek navigator, first sailed the open sea from Arabia to India and discovered the importance of the monsoon for navigation.
○ *remains of a crater (58 km/1 230 m). Rima Hippalus I is on its floor*

HUMORUM, MARE Sea of Moisture.
○ *circular lunar mare with a surface area of 117 000 sq km; it is only a little larger than Iceland*

KELVIN, PROM. Cape Kelvin. (33W, 27S) William Thomson, Lord Kelvin, 1824—1907; English physicist; worked in the field of thermodynamics, electricity; made over 60 inventions, constructed submarine cable.

LOEWY (33W, 23S) Moritz Loewy, 1833—1907; French astronomer, Director of Paris Observatory; designed the telescopic system 'equatorial coudé'; worked in the field of astrometry; made a map of the Moon.
○ *flooded crater (22×26 km/1 090 m)*

PUISEUX (39W, 28S) Pierre Puiseux, 1855—1928; French astronomer; made over 6 000 photographs of the Moon using the 'equatorial coudé' (see Loewy); co-author of the famous 'Atlas of the Moon'. ○ *flooded crater (25 km/400 m)*

(Herigonius) 41 52

Gassendi

Agatharchides

MARE

Loewy

Hippalus

HUMORUM

Puiseux

Pr. Kelvin

Doppelmayer

Vitello 62

GASSENDI

53 BULLIALDUS

The western part of Mare Nubium with the prominent Bullialdus, which is one of the most attractive lunar craters. Other numerous and interesting objects include a 'bridge' across the valley Bullialdus W close to the crater Agatharchides O, a typical lunar dome Kies Pi, and the prominent rilles Rimae Hippalus and Rima Hesiodus.

BULLIALDUS (22W, 21S) Ismael Boulliaud, 1605—94; French astronomer, historian and theologian.
○ *very prominent crater (59 km/3 510 m) with terrace-like walls, central peaks and an interesting radial structure on the outside of the crater*

CAMPANUS (28W, 28S) Giovanni Campano, 13th century; Italian theologian, astronomer and astrologer.
○ *crater (48 km/2 080 m)*

DARNEY, see map 42.

EPIDEMIARUM, PALUS Marsh of Epidemics. Named by Schmidt.

GOULD (17W, 19S) Benjamin Apthorp Gould, 1824—96; American astronomer; founder of the 'Astronomical Journal'; telegraphic determination of longitude.
○ *remains of a crater (34 km)*

HIPPALUS, see map 52.

KIES (23W, 26S) Johann Kies, 1713—81; German mathematician and astronomer. ○ *flooded crater (44 km/380 m)*

KÖNIG (25W, 26S) Rudolf König, 1865—1927; Austrian selenographer, musician, merchant; built his own observatory; made 47 000 measurements of lunar formations; König's telescope Zeiss is still in use in the Prague Observatory.
○ *crater (22 km/2 440 m)*

LUBINIEZKY (24W, 18S) Stanislaus Lubiniezky, 1623—75; Polish astronomer; studied mainly comets.
○ *flooded crater (44 km/770 m)*

MERCATOR (26W, 29S) Gerard de Kremer, 1512—94; Belgian cartographer; originated 'Mercator's projection' frequently used on naval and astronomical maps.
○ *crater with a flooded floor (47 km/1 760 m)*

NUBIUM, MARE Sea of Clouds, see map 54.

OPELT (18W, 16S) Friedrich W. Opelt, 1794—1863; German financier; patron of selenographers Lohrmann and Schmidt.
○ *remains of a crater (51 km)*

Small craters: Opelt E (8.0 km/1 370 m)
Kies E (6.5 km/1 120 m)

BULLIALDUS

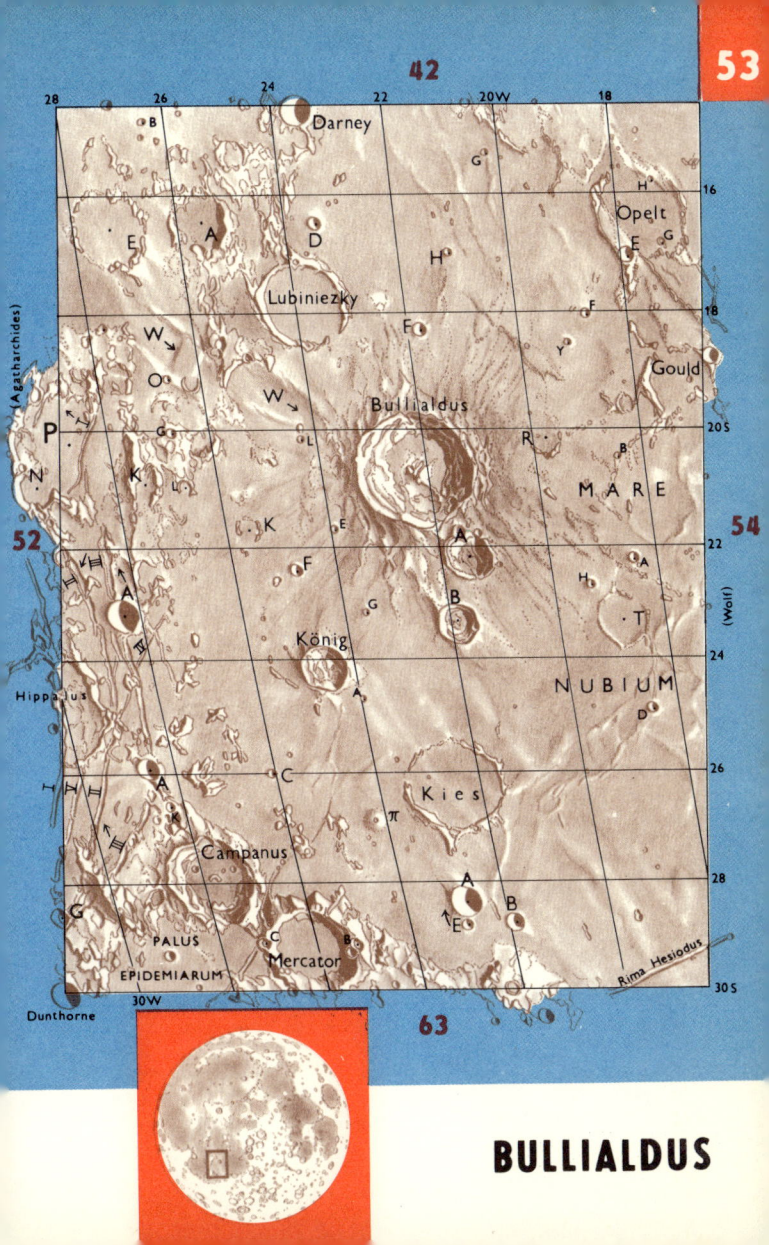

54 BIRT

The eastern part of Mare Nubium with many wrinkle ridges. Rupes Recta, the 'Straight Wall', is not far from the crater Birt; it is the most remarkable fault on the Moon. When illuminated from the east, it casts a wide shadow, which is clearly visible even through a small telescope. In the setting Sun it resembles a fine white line. Larger telescopes will reveal a narrow cleft between the craters Birt E (4.9 ×2.9km/600m) and Birt F (3.1km/470m).

BIRT (8W, 22S) William Radcliffe Birt, 1804—81; English astronomer, selenographer. ○ *crater (17 km/3 470 m).*
Birt A is situated at the edge of the wall (6.8 km/1 040 m)

HESIODUS (16 W, 29 S) About 735 BC; Greek poet.
○ *flooded crater (42 km); Hesiodus A has double concentric walls*

LASSELL (8W, 15S) William Lassell, 1799—1880; English amateur astronomer; with his home-made telescope discovered 4 satellites of the following planets: 1 of Neptune, 1 of Saturn (independently of W. C. Bond), 2 of Uranus. He also discovered 600 nebulae in the course of two years. ○ *crater (23 km/910 m); Lassell D (1.7 km/400 m) looks like a bright patch*

LIPPERSHEY (10W, 26S) Hans Lippershey (Jan Lapprey), died in 1619; Dutch spectacle-maker; reputed inventor of telescope.
○ *crater (6.8 km/1 350 m)*

NICOLLET (12W, 22S) Jean Nicholas Nicollet, 1788—1843; French selenographer. ○ *regular crater (15.2 km/2 030 m)*

NUBIUM, MARE Sea of Clouds. Its surface area is 264 000 sq km and has an approximately circular shape. Its northern border is not clearly defined.

PITATUS (14W, 30S) Pietro Pitati, 16th century; Italian mathematician and astronomer. ○ *flooded walled plain (105 km)*

RECTA, RUPES Straight Fault, previously called the 'Straight Wall'. Its length is 26 km, height 240 to 300 m and its apparent width is about 2.5 km. It is not a steep slope, as was considered in the past, but a moderate slope with a gradient of about 7° (1 : 9).

TAENARIUM, PROM. Cape Taenarium. Named by Hevelius after Cape Matapan (Tainaron) on Peloponnesus.

WOLF (17W, 23S) Max Wolf, 1863—1932; German astronomer; developed photographic method for finding small planets; with his assistants discovered over 300 plantes. ○ *remains of a flooded crater (25 km). Its wall si linked with crater Wolf B.*

BIRT

55 ARZACHEL

The area surrounding the prime meridian south of the ringed plains Ptolemaeus and Alphonsus, which together with crater Arzachel form a characteristic trio, is a region of large craters.

ALIACENSIS, see map 65.
ALPETRAGIUS (4W, 16S) Nur ed-din al Betrugi, 12th century; Arabian astronomer. ○ *crater (40 km/3 900 m)*
ALPHONSUS, see map 44.
ARZACHEL (2W, 18S) Al Zarkala, about 1028−87; Spanish-Arabian astronomer, author of 'Toledo tables'. ○ *very prominent crater (97 km/3 610 m), terraces, clefts in its floor. Formations E and F are valleys, parallel with the ridge of the wall.*
BLANCHINUS (2E, 25S) Giovanni Bianchini, about 1458; Italian teacher of astronomy. ○ *crater (58 × 68 km)*
DELAUNAY (2E, 22S) Charles Eugène Delaunay, 1816−72; French astronomer. ○ *formation 50 km in diameter, heart-shaped, divided by the central mountain range*
DONATI (5E, 21S) Giovanni Battista Donati, 1826−73; Italian astronomer; discovered 7 comets ('Donati's comet' in 1858). ○ *crater (36 km) with a central peak*
FAYE (4E, 21S) Hervé Faye, 1814−1902; French astronomer; discovered 'Faye's comet' in 1843. ○ *considerably disintegrated crater (37 km) with a central peak*
KRUSENSTERN (6E, 26S) Adam Johann, Baron von Krusenstern, 1770−1846; Russian admiral; circumnavigated the world in 1803−6. ○ *crater (47 km) with a flat floor*
LA CAILLE (1E, 24S) Nicholas Louis de la Caille, 1713−62; French astronomer. ○ *flooded crater (68 km)*
PARROT (3E, 15S) Johann J. F. W. Parrot, 1792−1840; German physicist and explorer. ○ *remains of a walled plain (68 km)*
PURBACH (2W, 26S) Georg von Peuerbach, 1423−61; German astronomer. ○ *walled plain (118 km/2 980 m)*
REGIOMONTANUS (1W, 28S) Johann Müller, 1436−76; German astronomer; critically assessed Ptolemy's 'Almagest'. ○ *irregular walled plain (126 × 110 km/1 730 m). Central peak with crater A (5.6 km/1 200 m)*
THEBIT (4W, 22S) Thebit ben Korra, 826−901; Baghdad astronomer; translated the 'Almagest' into Arabic. ○ *crater (55 km/3 270 m) Thebit A (20 km/2 720 m)*
WERNER (3E, 28S) Johan Werner, 1468−1528; German astronomer. ○ *prominent crater, terraces (70 km/4 220 m)*

ARZACHEL

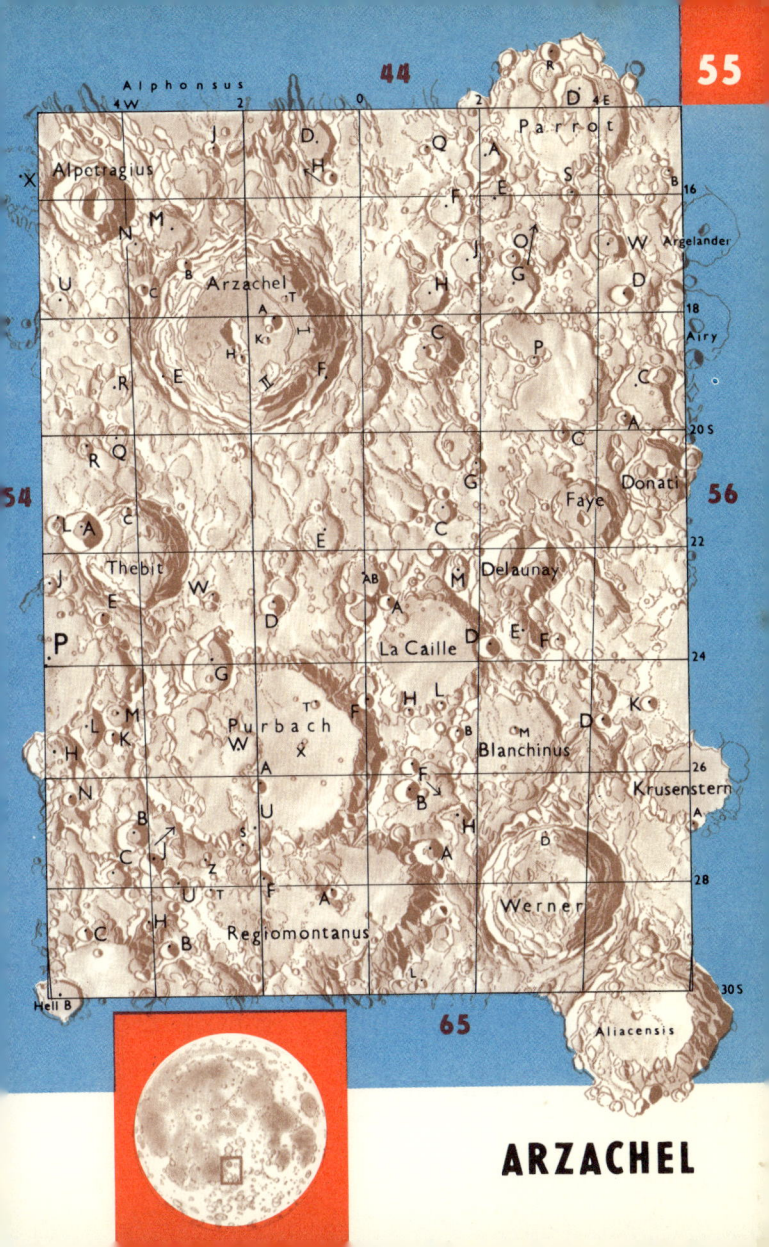

56 AZOPHI

A continental region with a number of craters, which make orientation very difficult. However, the pair Azophi and Abenezra and the prominent craters Almanon, Geber, Playfair and Apianus are useful points of orientation. Also conspicuous is the row of craters Airy, Argelander and Vogel.

ABENEZRA (12E, 21S) Abraham bar Rabbi ben Ezra, about 1092–1167; Jewish scholar, theologian, philosopher, mathematician and astronomer.
○ *polygonal crater (42 km/3 730 m)*
AIRY (6E, 18S) George Biddell Airy, 1801–92; English astronomer, Astronomer Royal.
○ *crater with a greatly articulated wall (37 km)*
ALMANON (15E, 17S) Abdalla Al Mamun, died in 833; Calif from Baghdad, son of Harun al-Raschid, patron of sciences.
○ *crater (49 km/2 480 m)*
APIANUS (8E, 27S) Peter Bienewitz, 1495–1552; German mathematician and astronomer; author of 'Astronomicum Caesareum'.
○ *crater (66 km/2 080 m)*
ARGELANDER (6E, 17S) Friedrich Wilhelm August Argelander, 1799–1875; German astronomer; author of 'Bonner Durchmusterung' — summary of 300 000 stars in the northern sky.
○ *crater (34 km/2 980 m)*
AZOPHI (13E, 22S) Abderrahman Al-Sufi, 903–986; Persian astronomer; compiled a stellar catalogue.
○ *crater (48 km/3 730 m)*
GEBER (14E, 20S) Gabir ben Aflah, died in about 1145; Spanish-Arabian astronomer. ○ *crater (46 km/3 510 m)*
KRUSENSTERN, see map 55.
PLAYFAIR (8E, 24S) John Playfair, 1748–1819; English mathematician and physicist. ○ *crater (48 km/2 910 m)*
PONTANUS (14E, 28S) Giovanni Gioviano Pontano, 1427–1503; Italian poet and astronomer. ○ *crater (54 km)*
SACROBOSCO (17E, 24S) John Holywood, died in 1256; English teacher of mathematics. ○ *crater (96 km)*
VOGEL (6E, 15S) Hermann Karl Vogel, 1841–1907; German astrophysicist; applied spectroscopy, spectral classification of the stars. ○ *crater (27 km/2 780 m)*
Small craters: Sacrobosco A (17.7 km/1 830 m)
Sacrobosco B (14.4 km/1 210 m)
Sacrobosco C (13.4 km/2 630 m)

AZOPHI

57 CATHARINA

The Altai Range, brightly illuminated by light coming from the east, crosses this region from crater Tacitus to the south and south-east. A bright, wide ray radiating from the crater Tycho (map 64) runs over the craters Polybius A and Polybius B. The large crater Catharina is a conspicuous object together with the craters Cyrillus and Theophilus (map 46).

ALTAI, RUPES Altai Range. A mountain range on the perimeter of the basin of Mare Nectaris, named by Mädler. The range resembles a fault sloping to the basin.

CATHARINA (24E, 18S) St Catharina of Alexandria, died in 307 AD; patron of Christian philosophers.
○ *considerably disintegrated circular mountain range (97 km/ 3 130 m)*

FERMAT (20E, 23S) Pierre de Fermat, 1601–65; French scholar, mathematician; made discoveries in the theory of numbers. ○ *crater (38 km)*

POLYBIUS (26E, 22S) About 204–122 BC; Greek historian and statesman; friend of Scipion from Africa.
○ *crater (42 km/2 050 m)*

PONS (22E, 25S) Jean Louis Pons, 1761–1831; French astronomer; discovered 30 comets. ○ *irregular crater (44 × 31 km)*

TACITUS (19E, 16S) Cornelius Tacitus, about 55–120 AD; Roman historian, author of the 'Life of Agricola', 'Germania' etc. ○ *prominent crater (40 km/2 840 m)*

WILKINS (20E, 30S) Hugh P. Wilkins, 1896–1960; English selenographer, author of very detailed maps of the Moon.
○ *considerably disintegrated, flooded crater (59 km)*

Small craters: Polybius A (16.8 km/3 720 m)
　　　　　　　Pons B (13.9 km/3 050 m)
　　　　　　　Polybius B (12.8 km/2 630 m)
　　　　　　　Tacitus N (7.1 km/1050 m)

CATHARINA

58 FRACASTORIUS

The dark surface of Mare Nectaris, the flooded walled plain Fracastorius and the beautiful crater Piccolomini are the most prominent formations in this part of the Moon. A low mountain range runs north of the crater Beaumont.

BEAUMONT (29E, 18S) Elie de Beaumont, 1798—1874; French geologist; developed von Buch's theory of the evolution of mountain chains and demonstrated the method of determining the relative age of individual layers.
○ *crater (53 km); its wall is interrupted in the east.*
BOHNENBERGER (40E, 16S) Johann Gottlieb Friedrich von Bohnenberger, 1765—1831; German mathematician and astronomer. ○ *crater (33 km/1 060 m), uneven floor with hills*
FRACASTORIUS (33E, 21S) Girolamo Fracastoro, 1483—1553; Italian physicist, astronomer and poet; in his 'Homocentrica' he attempted to replace Ptolemy's system with an inflexible system of homocentric spheres.
○ *walled plain (124 km), wall in the north is missing and its floor continues into M. Nectaris*
NECTARIS, MARE Sea of Nectar. Circular mare with a surface area of 100 000 sq km. Flooded central part of lunar basin the outer wall of which follows the Altai Range.
PICCOLOMINI (32E, 30S) Alessandro Piccolomini, 1508—78; Italian archbishop and astronomer; made stellar maps in which the stars were marked for the first time with letters of the Latin alphabet; Bayer's system using Greek letters was accepted at a later date.
○ *very prominent crater (89 km); central mountain-massif*
PYRENAEUS, MONTES Pyrenees, see map 48.
ROSSE (35E, 18S) William Parsons, Earl of Rosse, 1800—67; Irish nobleman, astronomer; erected a reflecting telescope having a mirror 180 cm in diameter in Parsonstown; studied the details of nebulae and the spiral structure of galaxies; named the following nebulae: Owl Nebula, Crab Nebula, Dumb-bell Nebula and others. ○ *crater (12 km/2 420 m)*
WEINEK (37E, 28S) Ladislav Weinek, 1848—1913; Austrian astronomer; from 1883 Director of Prague Observatory at Klementinum; prepared an atlas of the Moon.
○ *crater (32 km/3 370 m)*

MARE NECTARIS

Beaumont · Rosse · Bohnenberger · Fracastorius · Santbech · Weinek · (Neander) · Piccolomini · (Reichenbach)

FRACASTORIUS

59 PETAVIUS

The southern promontory of Mare Fecunditatis with bright rays radiating from craters Petavius B and Snellius A, as well as the giant crater Petavius and a group of craters named after famous navigators, are the most prominent features of this region.

BIOT (51E, 23S) Jean-Baptiste Biot, 1774—1862; French astronomer, geodesist, historian of astronomy. ○ *crater (13 km)*

BORDA (47E, 25S) Jean Charles Borda, 1733—99; French naval officer and astronomer.
○ *crater (44 km), disintegrated wall, central peak*

COLOMBO (46E, 15S) Cristoforo Colombo (Columbus), about 1446—1506; Genoese navigator, discovered America (1492).
○ *prominent crater (78 km) central peaks*

COOK (49E, 18S) James Cook, 1728—79; English naval captain and explorer, twice circumnavigated the world.
○ *flooded crater with a low wall (47 km)*

FECUNDITATIS, MARE Sea of Fertility, see map 48.

HASE (63E, 29S) Johann M. Hase, 1684—1742; German mathematician and cartographer. ○ *disintegrated crater (83 km)*

MCCLURE (50E, 15S) Robert le Mesurier McClure, 1807—73; British naval officer, found the north-western passage.
○ *crater (24 km)*

MONGE (48E, 19S) Gaspard Monge, 1746—1818; French mathematician; laid foundations of modern geometry.
○ *crater (37 km)*

PALITZSCH (64E, 28S) Johann G. Palitzsch, 1723—88; German amateur astronomer; first to find Halley's comet in 1758.
○ *inconspicuous crater (41 km)*

PETAVIUS (60E, 25S) Denis Petau, 1583—1652; French theologian and historian; studies in chronology.
○ *crater (177 km); the floor of the central mountain chain has clefts and dark patches*

SANTBECH (44E, 21S) Daniel Santbech Noviomagus, about 1561; Dutch mathematician and astronomer.
○ *crater (64 km)*

SNELLIUS (56E, 29S) Willibrord Snell, 1591—1626; Dutch astronomer and geodesist; discovered law of refraction of light.
○ *crater (83 km)*

WROTTESLEY (57E, 24S) John, Baron Wrottesley, 1798—1867; English amateur astronomer; worked in the field of astrometry, catalogue of double stars.
○ *prominent crater (57 km)*

PETAVIUS

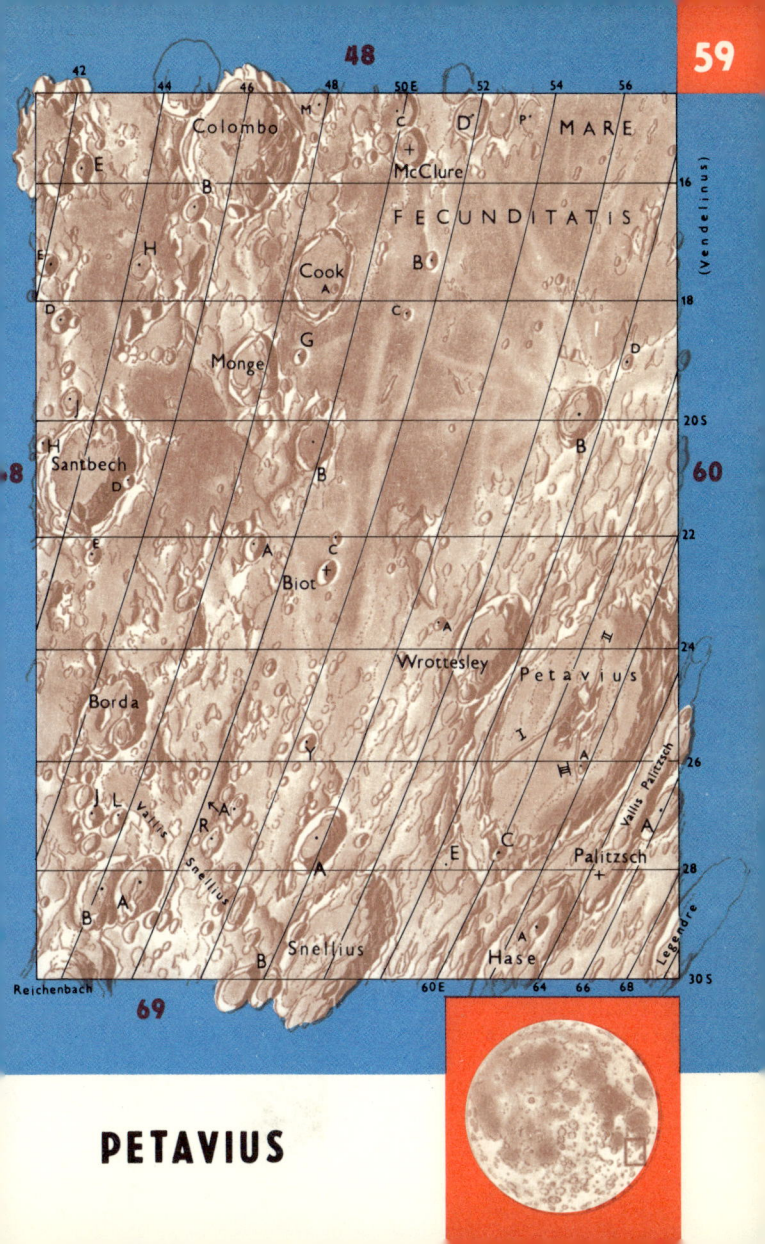

Map labels:
48, 59, 42, 44, 46, 48, 50 E, 52, 54, 56
Colombo, M, C, D, P, MARE
E, B, McClure, 16
F, H, FECUNDITATIS, (Vendelinus)
Cook, B
D, A, C
G, D
Monge
J, 20 S
H, Santbech, B, 60
D
E, A, C, 22
Biot, F
A
Wrottesley, Petavius, 24
Borda, I, A
Y, 26
I, L, Vallis, A, R, Vallis Palitzsch, A
Snellius, A, E, C, Palitzsch, 28
B, A, A, Legendre
Reichenbach, Snellius, Hase, 30 S
69, 60 E, 64, 66, 68

60 VENDELINUS

The eastern limb of the near side of the Moon contains the large walled plain Humboldt, which is best seen shortly after Full Moon. The walled plain Vendelinus belongs to the well-known line of craters Langrenus, Vendelinus, Petavius and Furnerius.

BALMER (70E, 20S) Johann Jakob Balmer, 1825—98; Swiss mathematician and physicist; formulated rule for wavelength of lines in the spectrum of hydrogen ('Balmer's series').
○ *remains of a flooded walled plain (112 km)*
BEHAIM (80E, 17S) Martin Behaim (Behem), 1436—1506; German navigator and cartographer.
○ *regular crater with terrace-like walls and a central peak (56 km)*
GIBBS (84E, 18S) Josiah Willard Gibbs, 1839—1903; American mathematician and physicist. ○ *crater (81 km)*
HECATAEUS (79E, 22S) Died about 476 BC; Greek historian of Miletus, author of a description of the world with a map.
○ *walled plain (127 km)*
HOLDEN (62E, 19S) Edward Singleton Holden, 1846—1914; American astronomer; first Director of Lick Observatory.
○ *crater (47 km)*
HUMBOLDT (81E, 27S) Wilhelm von Humboldt, 1767—1835; brother of Alexander von Humboldt, German statesman and philologist.
○ *typical walled plain (201 km); its floor has a central mountain range, a network of concentric and radial clefts and dark patches close to the wall*
LAMÉ (64E, 15S) Gabriel Lamé, 1795—1870; French mathematician. ○ *crater (84 km)*
LEGENDRE (70E, 29S) Adrien Marie Legendre, 1752—1833; French mathematician; known for his work on elliptic integrals, theory of numbers. ○ *walled plain (79 km)*
PHILLIPS (76E, 27S) John Phillips, 1800—74; English geologist, popularizer of sciences; observed Mars and the Moon.
○ *walled plain (101 km)*
SCHORR (90E, 19S) Richard Schorr, 1867—1951; German astronomer. ○ *crater (54 km)*
VENDELINUS (62E, 16S) Godefroid Wendelin, 1580—1667 Flemish astronomer. ○ *walled plain (147 km)*

VENDELINUS

(Kapteyn) 49

60

59

69

Lamé
H G M B
Vendelinus   B Behaim S
L Z N J
E P
F   Gibbs
W T   C B L K Schorr
Holden   Balmer
        Hecataeus
B
D W
        Phillips
B   Humboldt
Legendre L
K   Barnard
D A

(Palitzsch)

61 PIAZZI

The south-western limb of the Moon. This area is intersected by a rich system of radial mountain ranges and valleys, which radiate from the centre of Mare Orientale, e.g. Vallis Bouvard, Vallis Inghirami and Vallis Baade. However, this structure can be distinguished from the Earth only with great difficulty.

BAADE (82W, 44S) Walter Baade, 1893–1960; German astronomer; made significant contributions to an understanding of our galaxy, as well as other galaxies. ○ *crater (55 km)*

BOUVARD, VALLIS Bouvard's Valley (82W, 40S) Alexis Bouvard, 1767–1843; French mathematician and astronomer; discovered several comets.
○ *valley about 180 km long and 40 km wide*

CATALÁN (87W, 46S) Miguel A. Catalán, 1894–1957; Spanish physicist; worked in the field of spectroscopy.
○ *crater (14 km)*

FOURIER (53W, 30S) Jean Baptiste Joseph Fourier, 1768–1830; French physicist and mathematician; developed the mathematical series which bears his name. ○ *crater (51 km)*

GRAFF (88W, 43S) Kasimir Romuald Graff, 1878–1950; Polish astronomer and geodesist; Director of Vienna Observatory.
○ *crater (40 km)*

LACROIX (59W, 38S) Sylvestre François Lacroix, 1765–1843; French mathematician, teacher. ○ *crater (36 km)*

LAGRANGE (72W, 33S) Joseph Louis Lagrange, 1736–1813; prominent French mathematician; author of 'Mécanique Analytique'. ○ *disintegrated walled plain (160 km)*

LEHMANN, see map 62.

PIAZZI (68W, 36S) Guiseppe Piazzi, 1749–1826; Italian astronomer; discovered the first asteroid (Ceres).
○ *disintegrated walled plain (101 km)*

SHALER (85W, 33S) Nathaniel Southgate Shaler, 1841–1906; American geologist and palaeontologist; interpreted the photographs of the Moon in geological terms. ○ *crater (44 km)*

WRIGHT (87W, 31S) 1. Frederick E. Wright, 1878–1953; American astronomer, selenologist. 2. Thomas Wright, 1711–1786; English natural philosopher. 3. William Hammond Wright, 1871–1959; American astronomer; photographed Mars. ○ *crater (36 km)*

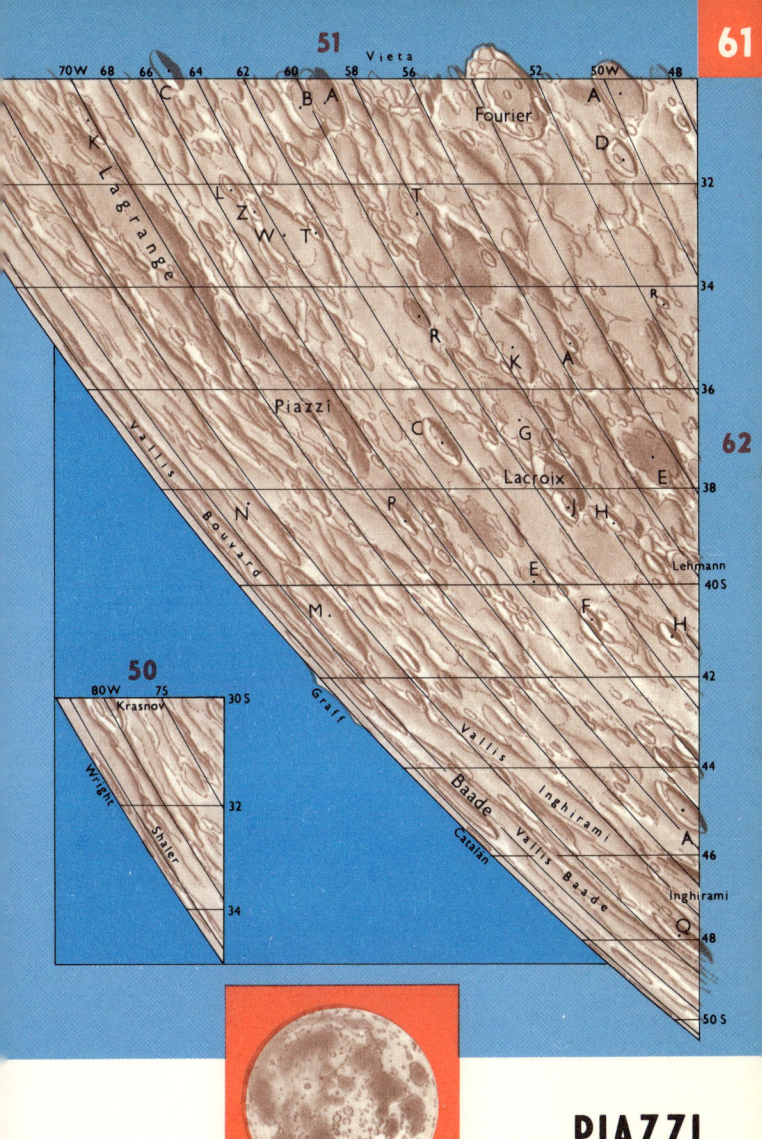

PIAZZI

62 SCHICKARD

Mountainous region along the south-western limb of the Moon, which in the north reaches the edge of Mare Humorum. One of the largest walled plains, Schickard, with characteristic dark patches on its floor, lies in this area. Close to the crater Clausius there is a small, unnamed mare.

CLAUSIUS (44W, 37S) Rudolf Julius Emmanuel Clausius, 1822—88; German physicist; worked in the field of thermodynamics and kinetic theory of gases.
○ *crater with a flooded floor (22 km)*

DREBBEL (49W, 41S) Cornelius Drebbel, 1572—1634; Dutch inventor; claimed invention of the telescope and the microscope.
○ *crater (30 km)*

INGHIRAMI (69W, 48S) Giovanni Inghirami, 1779—1851; Italian astronomer.
○ *prominent crater (91 km), crossed by radial mountain ranges and valleys running from the centre of Mare Orientale*

LEE (41W, 31S) John Lee, 1783—1866; English collector of antiquities; selenographer.
○ *remains of a flooded crater (42 km/1 340 m)*

LEHMANN (56W, 40S) Jacob H. W. Lehmann, 1800—63; German theologian and astronomer; worked in the field of celestial mechanics. ○ *considerably eroded crater (53 km)*

LEPAUTE (34W, 33S) Mme Lepaute, née Nicole Reine Etable de la Brière, 1723—88; French arithmetician, cooperated with Clairaut and Lalande. ○ *crater (15 km/2 070 m)*

SCHICKARD (55W, 44S) Wilhelm Schickard, 1592—1635; German mathematician and astronomer; first to attempt determination of the orbit of a meteor by simultaneous observations at different places.
○ *vast walled plain (227 km) with partially flooded floor*

VITELLO (38W, 30S) (or Witelo), about 1270; Polish philosopher; worked on optics. ○ *crater (45 km/1 730 m)*

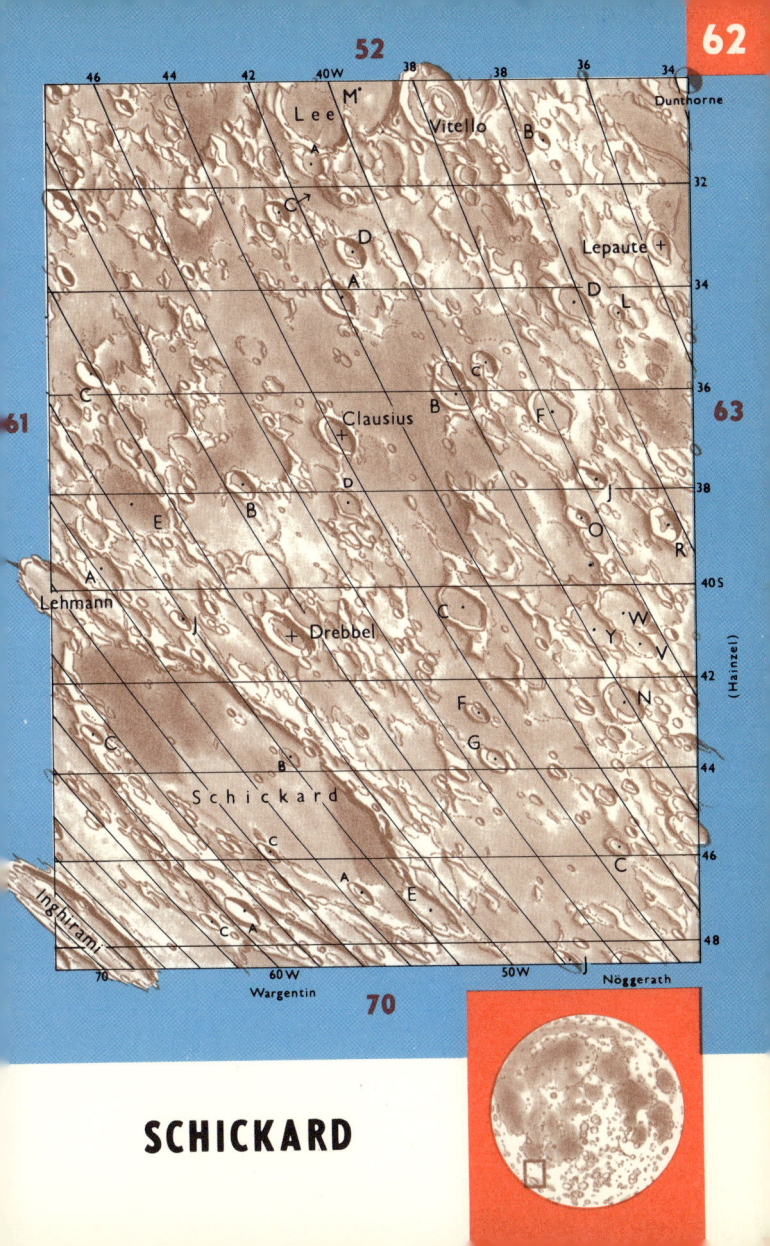

SCHICKARD

Lee M
Vitello
Dunthorne
Lepaute
Clausius
Lehmann
Drebbel
Schickard
Inghirami
Wargentin
Nöggerath
(Hainzel)

63 CAPUANUS

The area in the south-western section of the Moon. On the surface of Palus Epidemiarum there is a system of clefts close to the crater Ramsden and the wide Rima Hesiodus (which continues on map 53). The floor of the crater Capuanus contains a group of domes.

CAPUANUS (27W, 34S) Francesco Capuano di Manfredonia, 15th century; Italian theologian and astronomer.
○ *flooded crater (60 km)*
CICHUS (21W, 33S) Francesco de gli Stabilli, also known as Cecco d'Ascoli, 1257—1327; Italian astronomer and astrologer; accused of heresy and burnt in Florence. ○ *crater (40 km/ 2 760 m); on its western wall is Cichus C (11.1 km/1250 m)*
DUNTHORNE (32W, 30S) Richard Dunthorne, 1711—75; English geodesist, astronomer. ○ *crater (16 km/2 780 m)*
ELGER (30W, 35S) Thomas G. Elger, 1838—97; English selenographer; produced a map of the Moon (1895).
○ *crater (20 km/1 250 m)*
EPIMENIDES (30 W, 41 S) About 596 BC; Cretan poet and augur.
○ *crater (27 km)*
HAIDINGER (25W, 39S) Wilhelm Karl von Haidinger, 1795—1871; Austrian geologist and physicist.
○ *crater (21 km/2 330 m); Haidinger B is situated on the south-eastern wall (10.3 km/1 500 m)*
HAINZEL (34W, 41S) Paul Hainzel, about 1570; German astronomer. ○ *complex formation consisting of three intersecting craters, the largest (Hainzel) is 70 km in diameter and two smaller ones are marked A and C*
LAGALLA (22W, 45S) Giulio Cesare Lagalla, 1571—1624; Italian philosopher, one of the earliest observers of the Moon with the telescope. ○ *remains of a crater (85 km)*
MARTH (29W, 31S) Albert Marth, 1828—97; German astronomer. ○ *interesting crater with double wall, 6.1 km in diameter*
MEE (35W, 44S) Arthur Butler P. Mee, 1860—1926; Scottish astronomer; observed the Moon and Mars.
○ *disintegrated crater (132 km)*
RAMSDEN (32W, 33S) Jesse Ramsden, 1735—1800; English mechanician. ○ *flooded crater (24 km/1 990 m)*
WEISS (20W, 32S) Edmund Weiss, 1837—1917; Austrian astronomer; co-founder and Director of the Observatory in Vienna; determined orbits of celestial bodies and the origin of Leonids. ○ *remains of a flooded crater (63 km)*

CAPUANUS

64 TYCHO

Crater field south of Mare Nubium. This area is dominated by the crater Tycho, which is one of the most prominent craters on the Moon and the centre of the most extensive system of bright rays.

BALL (8W, 36S) William Ball, died in 1690; English amateur astronomer; confirmed the observations of Huygens on Saturn's rings. ○ *crater (40 km/2 810 m)*

BROWN (18W, 46S) Ernest William Brown, 1866−1938; Anglo-American astronomer; theory of the motion of the Moon. ○ *crater (30 km); its wall is penetrated by the crater Brown E*

DESLANDRES, see map 65.

GAURICUS (12W, 34S) Luca Gaurico, 1476−1558; Italian theologian, astronomer and astrologer; translator of the 'Almagest'. ○ *considerably eroded crater (79 km)*

HEINSIUS (18W, 40S) Gottfied Heinsius, 1709−69; German mathematician and astronomer. ○ *crater (66 km/2 650 m); merges with Heinsius A (20 km/3 270 m), B and C*

HELL (8W, 32S) Maximilian Hell, 1720−92; Hungarian astronomer; founder of Vienna Observatory; observed transit of Venus. ○ *crater (33 km/2 200 m)*

MONTANARI (21W, 46S) Geminiano Montanari, 1633−87; Italian astronomer; first micrometric measurements for the mapping of the Moon. ○ *disintegrated crater (77 km)*

PICTET (7W, 44S) Marc A. Pictet, 1752−1825; Swiss astronomer and naturalist. ○ *crater (62 km)*

PITATUS, see map 54.

SASSERIDES (9W, 39S) Gellio Sasceride, 1562−1612; Danish physician and astronomer; assistant to Tycho Brahe. ○ *considerably disintegrated crater (90 km)*

STREET (10W, 46S) Thomas Streete (or Street), about 1661; English astronomer, author of 'Astronomia Carolina'. ○ *crater (58 km)*

TYCHO (11W, 43S) Tycho Brahe, 1546−1601; Danish astronomer; prominent observer and organizer of scientific research; his accurate observations enabled Kepler to discover the laws of planetary motion. ○ *crater (85 km/4 850 m), vast terrace-like wall, central massif 1 600 m high*

WILHELM (21W, 43S) Wilhelm IV of Hessen, 1532−92; German statesman, astronomer. ○ *walled plain (107 km)*

WURZELBAUER (16W, 34S) Johann Philipp Wurzelbauer, 1651−1725; German astronomer, observer of the Sun. ○ *greatly disintegrated crater (82 km)*

64

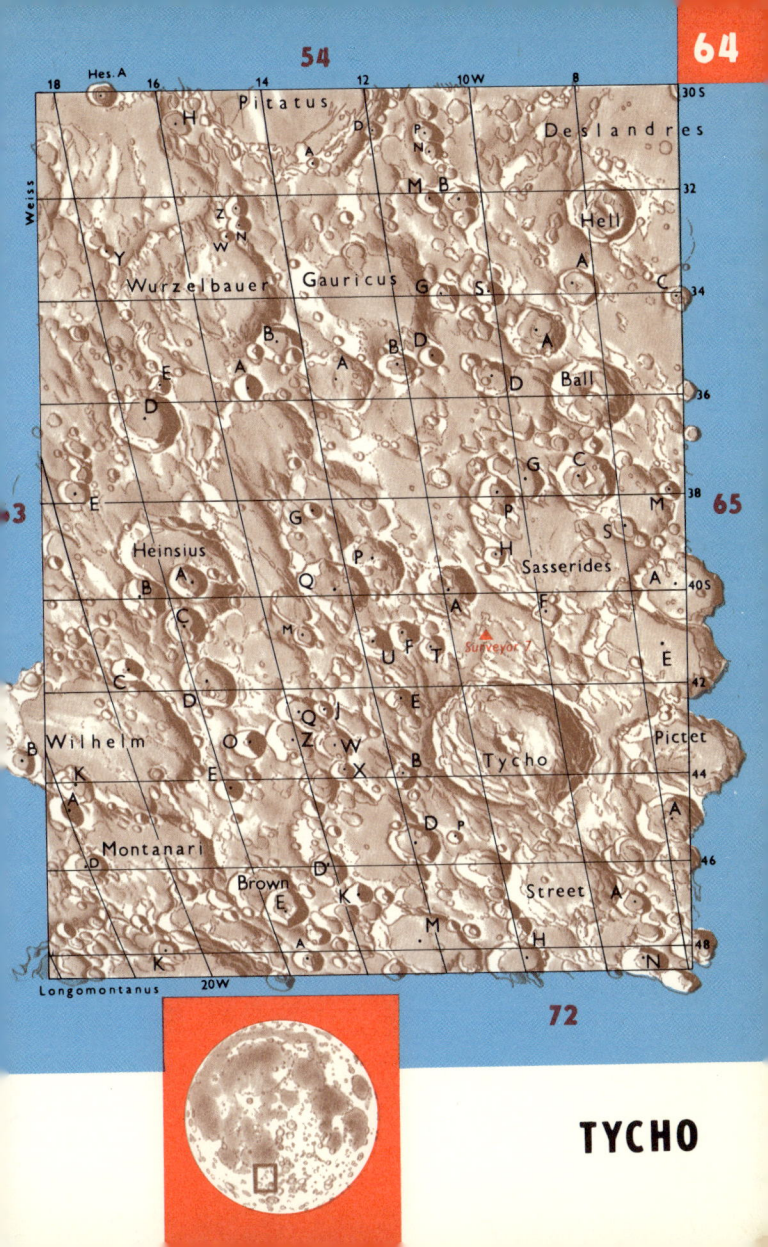

Pitatus — Deslandres — Hell — Wurzelbauer — Gauricus — Ball — Heinsius — Sasserides — Surveyor — Wilhelm — Tycho — Pictet — Montanari — Brown — Street — Longomontanus — Weiss — Hes. A

TYCHO

65 WALTER

Dense crater field in the vicinity of the prime meridian on the southern hemisphere of the Moon. The main points of orientation in this area are the walled plain Stöfler, the pair of craters Aliacensis and Werner (see also map 55) and a group of craters surrounding the crater Huggins.

ALIACENSIS (5E, 31S) Pierre d'Ailly, 1350—1420; French theologian and geographer. ○ *crater (80 km/3 680 m)*

DESLANDRES (5W, 32S) Henri Alexandre Deslandres, 1853—1948; French astronomer, observer of the Sun; invented spectroheliograph; Director of Observatory at Meudon.
○ *considerably disintegrated walled plain 234 km in diameter*

FERNELIUS (5E, 38S) Jean Fernel, 1497—1558; French physicist. ○ *crater with a flooded floor (70 km)*

HUGGINS (1W, 41S) Sir William Huggins, 1824—1910; English astronomer; pioneered astronomical spectroscopy.
○ *crater (63 km), merges with Nasireddin*

LEXELL (4W, 36S) Anders J. Lexell, 1740—84; Finnish astronomer; worked in the field of celestial mechanics.
○ *crater (63 km)*

LICETUS (7E, 47S) Fortunio Liceti, 1577—1657; Italian natural philosopher. ○ *crater (75 km)*

MILLER (1E, 39S) William A. Miller, 1817—70; English chemist. ○ *crater (61 km/3 550 m)*

NASIREDDIN (0.2E, 41S) Nasir-al-Din, 1201—74; Persian astronomer. ○ *crater (51 km)*

NONIUS (4E, 35S) Pedro Nuñez, 1492—1577; Portuguese mathematician; described a contrivance for graduating mathematical instruments, precursor of the Vernier.
○ *polygonal, disintegrated crater (70 km/2 990 m)*

ORONTIUS (4W, 40S) Orontius Finaeus, 1494—1555; French mathematician. ○ *walled plain (109 km)*

PICTET, see map 64.

PROCTOR (5W, 46S) Mary Proctor, 1862—1919; daughter of the astronomer R. A. Proctor; astronomer, popularizer of astronomy. ○ *crater (52 km)*

SAUSSURE (4W, 43S) Horace Benedict de Saussure, 1740—99; Swiss geologist, meteorologist. ○ *crater (56 km/1 880 m)*

STÖFLER (6E, 41S) Johann Stöffler, 1452—1534; German mathematician, astronomer and astrologer.
○ *walled plain with a flooded floor (137 km/2 760 m)*

WALTER (1E, 33S) Bernard Walter, 1430—1504; German astronomer. ○ *walled plain (132 × 140 km/4 130 m)*

55

4

6
4W
2
0
2
4E
6
32
34
36
38
40S
42
44
46
48
Deslandres
C
Aliacensis
B
D
A
W
E
Walter
L
K
Nonius
A
Lexell
P
R
B
D
S
H
A
Kaiser
B
E
A
Fernelius
C
C
A
F
D
Miller
K
Orontius
A
Huggins
Nasireddin
H
Stöfler
N
J
F
Pictet
Saussure
G
D
P
Faraday
A
E
E
C
J
E
F
H
Proctor
D
Licetus
U
A
C
K
P
D
Maginus

66

73

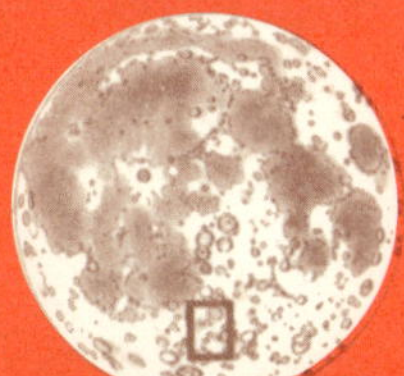

WALTER

66 MAUROLYCUS

Crater field in the south-eastern section of the southern hemisphere of the Moon. Prominent crater Maurolycus forms a characteristic pair with the neighbouring complex Faraday and Stöfler (see map 65). Crater Gemma Frisius is remarkable for its exceptionally high wall, which measures over 5 000 m.

BAROCIUS (17E, 45S) Francesco Barozzi, about 1570; Italian mathematician. ○ *crater (82 km); the wall is disintegrated in the north and intersected by craters B and C*

BREISLAK (18E, 48S) Scipione Breislak, 1748—1826; Italian geologist and chemist. ○ *crater (50 km)*

BUCH (18E, 39W) Christian Leopold von Buch, 1774—1853; German geologist. ○ *crater (54 km/1 440 m)*

BÜSCHING (20E, 38S) Anton F. Büsching, 1724—93; German geographer and philosopher. ○ *crater (52 km)*

CLAIRAUT (14E, 48S) Alexis Claude Clairaut, 1713—65; eminent French mathematician, geodesist and astronomer.
○ *crater (75 km); the southern part of the wall is intersected by craters A and B*

FARADAY (9E, 42S) Michael Faraday, 1791—1867; English chemist and physicist; known for his discoveries in electricity, magnetism etc. ○ *crater (69 km/4 090 m)*

GEMMA FRISIUS (13E, 34S) Reiner Jemma, 1508—55; Dutch geographer, cartographer and astronomer.
○ *crater with disintegrated wall (88 km/5 160 m); its adjacent craters are Gemma Frisius D (27.5 km/3 240 m) and Gemma Frisius EB (14.6 km/3 180 m)*

GOODACRE (14E, 33S) Walter Goodacre, 1856—1938; English selenographer; made a map of the Moon.
○ *crater (46 km/3 190 m)*

KAISER (7E, 36S) Frederick Kaiser, 1808—72; Dutch astronomer; observer of double stars and of Mars.
○ *crater (53 km); on its eastern wall is situated Kaiser A (21 × 14 km/2 330 m)*

MAUROLYCUS (14E, 42 S) Francesco Maurolico, 1494—1575; Italian mathematician, opponent of Copernicus' system.
○ *vast walled plain (114 km/4 730 m), central peaks*

POISSON (11E, 30S) Siméon Denis Poisson, 1781—1840; French mathematician, friend of Lagrange and Laplace; celestial mechanics. ○ *remains of a crater 44 km in diameter, which is connected by the walls of the neighbouring craters with crater Poisson T into a valley*

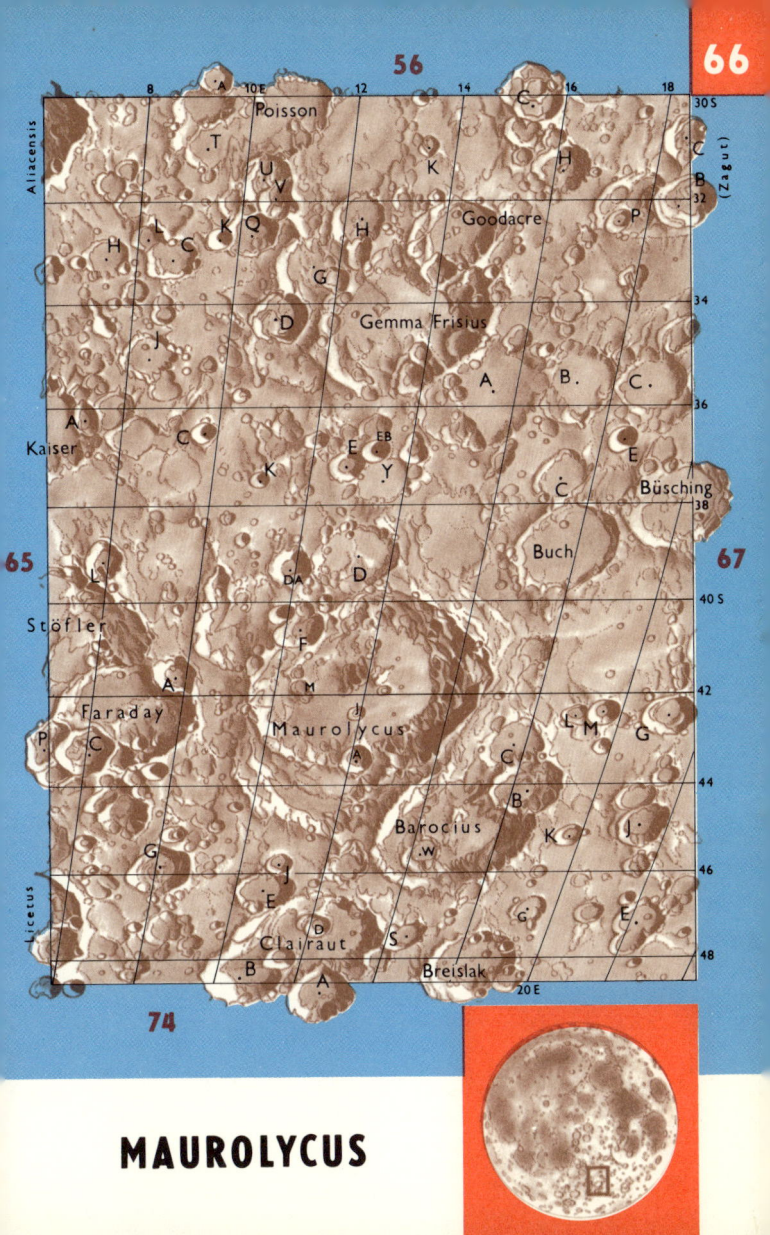

MAUROLYCUS

Poisson · Goodacre · Gemma Frisius · Kaiser · Büsching · Buch · Stöfler · Faraday · Maurolycus · Barocius · Clairaut · Breislak · Aliacensis · (Zagut) · Liceus

56 · 66 · 65 · 67 · 74

67 RABBI LEVI

Very dense and chaotic crater field in the south-eastern section of the southern hemisphere of the Moon. An important point of orientation is the crater Rabbi Levi, which has two pairs of small craters on its floor. At bottom left of the map is a part of the walled plain Janssen with a wide rille on its floor.

CELSIUS (20E, 34S) Anders Celsius, 1701—44; Swedish natural philosopher and astronomer; devised the Centigrade scale of temperature. ○ *crater (36 km)*

DOVE (32E, 47S) Heinrich Wilhelm Dove, 1803—79; German physicist; worked in the field of meteorology and electricity. ○ *crater (30 km)*

JANSSEN, see map 68.

LINDENAU (25E, 32S) Bernhard von Lindenau, 1780—1854; German astronomer, soldier and politician. ○ *prominent crater with terrace-like wall (53 km/2 930 m), central mountains*

LOCKYER (37E, 46S) Sir Joseph Norman Lockyer, 1836—1920; English astrophysicist; discovered helium in the Sun. ○ *crater (34 km) on the wall of Janssen*

NICOLAI (26E, 42S) Friedrich B. G. Nicolai, 1793—1846; German astronomer; orbits of comets, celestial mechanics. ○ *crater (42 km)*

RABBI LEVI (24E, 35S) Levi ben Gershon, 1288—1344; Spanish Jew, philosopher, mathematician, astronomer. ○ *crater (81 km); craters R. Levi L (12.6 km/2 410 m), R. Levi A (12.1 km/1 350 m) and others are situated on its floor*

RICCIUS (26E, 37S) Matteo Ricci, 1552—1610; Italian missionary in China; teacher of mathematics and astronomy. ○ *considerably disintegrated crater (71 km); south of this lies Riccius E (22 km/3 520 m)*

ROTHMANN (28E, 31S) Christopher Rothmann, died about 1600; German astronomer. ○ *crater (43 km/4 220 m)*

SPALLANZANI (25E, 46S) Lazaro Spallanzani, 1729—99; Italian scientist, physiologist, traveller. ○ *crater (32 km)*

STIBORIUS (32E, 34S) Andreas Stoberl, 1465—1515; Austrian philosopher, theologian, astronomer. ○ *crater (43 km/3750m)*

WÖHLER (32E, 38S) Friedrich Wöhler, 1800—82; German chemist; isolated aluminium and beryllium. ○ *crater (26 km/2 050 m)*

ZAGUT (22E, 32S) Abraham ben Samuel Zaguth, end of 15th century; Spanish Jew, astronomer, astrologer. ○ *crater (84 km); Zagut B (32 km/3 410 m) is situated west of it*

RABBI LEVI

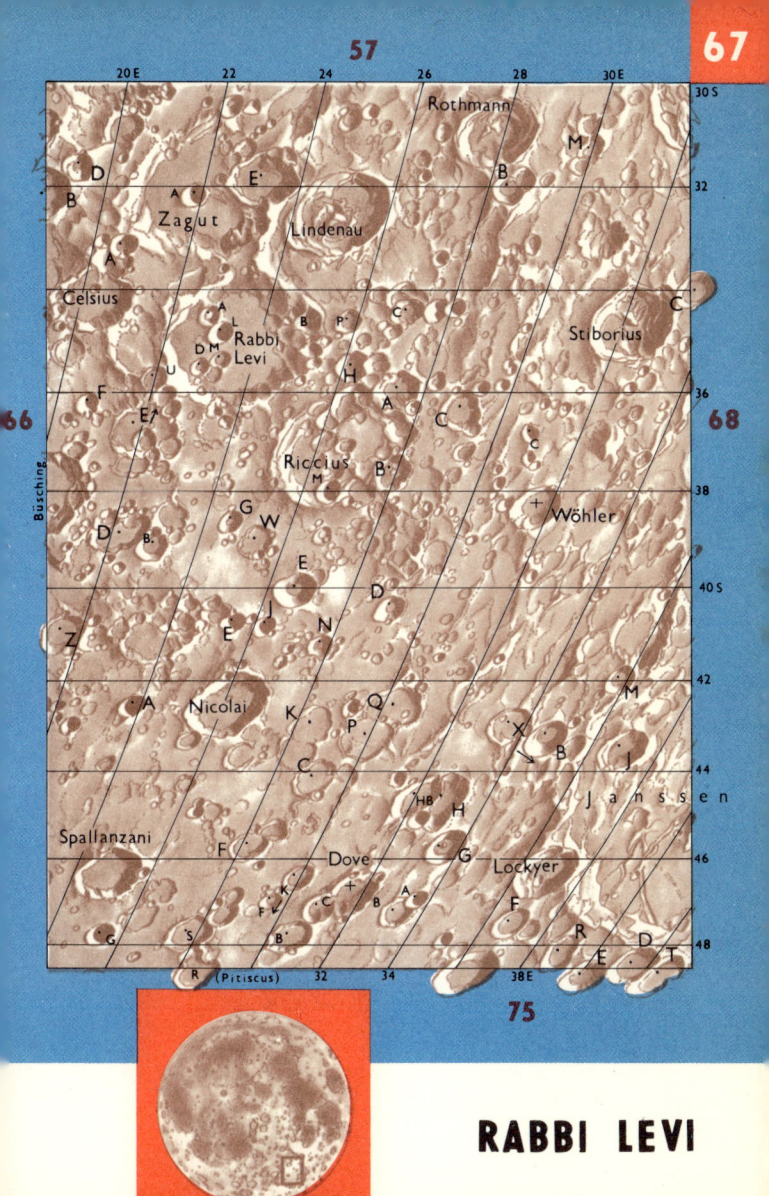

68 RHEITA

The crater field situated along the south-eastern limb of the Moon. It is intersected by a system of faults and valleys, which originate in the centre of Mare Nectaris, of which the long valley of Rheita is the most prominent.

BRENNER (39E, 39S) Leo Brenner, 1855—1928; Austrian amateur astronomer; observer of the Moon and the planets.
○ *greatly eroded crater (87 km)*

FABRICIUS (42E, 43S) David Goldschmidt, 1564—1617; Dutch amateur astronomer. ○ *prominent crater (78 km)*

JANSSEN (42E, 45S) Pierre Jules César Janssen, 1824—1907; French astronomer, Director of Observatory at Meudon from 1875. ○ *walled plain (190 km); its floor has a wide cleft, mountainous massif etc.*

MALLET (54E, 45S) Robert Mallet, 1810—81; Irish civil engineer and geologist.
○ *crater (58 km) adjacent to Vallis Rheita*

METIUS (43E, 40S) Adriaan Adriaanszoon Metius, 1571—1635; Dutch mathematician and astronomer. ○ *crater (88 km)*

NEANDER (40E, 31S) Michael Neumann, 1529—1581; German mathematician, physician and astronomer.
○ *crater (52 km/3 400 m), central mountains*

PEIRESCIUS (68E, 46S) Nicolas Claude Fabri de Peiresc, 1580—1637; French natural philosopher and amateur astronomer; discovered the Great Nebula in Orion in 1610.
○ *crater (62 km)*

REIMARUS (60E, 48S) Nicolai Reymers Bär (Ursus), died in 1600; German mathematician; was charged with plagiarism as a result of publishing a description of the planetary system very similar to that of Tycho Brahe. ○ *crater (48 km)*

RHEITA (47E, 37S) Anton Maria Schyrleüs (Širek) of Rheita, 1597—1660; Czech optician; made Kepler's telescope; astronomer; prepared a map of the Moon.
○ *crater (70 km) on the northern edge of Vallis Rheita. Crater valley Rheita E (66 × 32 km)*

STEINHEIL, see map 76.

VEGA (63E, 45S) Georg, Freiherr von Vega, 1756—1802; German mathematician; computed accurate logarithmic tables.
○ *crater (76 km)*

YOUNG (51E, 42S) Thomas Young, 1773—1829; English physician and natural philosopher; established the wave theory of light. ○ *crater (72 km)*

RHEITA

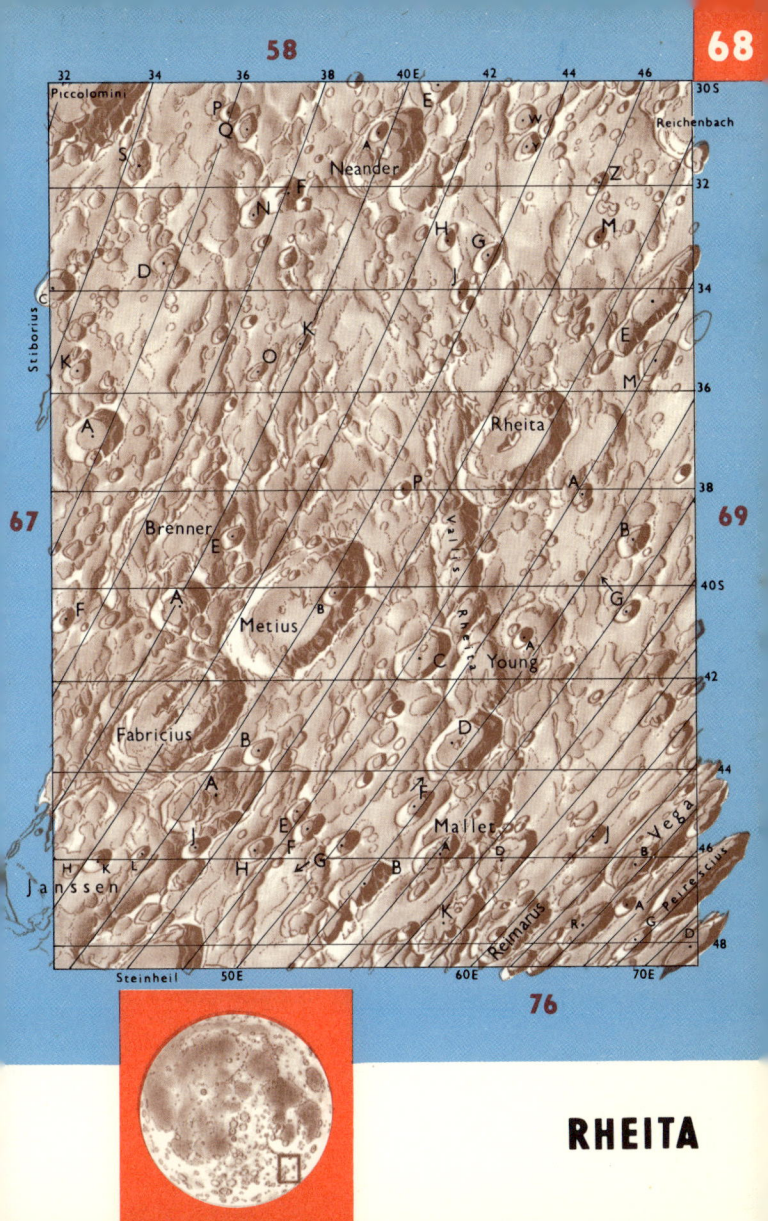

69 FURNERIUS

The south-eastern limb of the Moon. Useful points of orientation are the craters Stevinus, Furnerius and Oken with a dark floor, as well as a part of Mare Australe, which stretches from the south.

ABEL (85E, 34S) Niels Henrik Abel, 1802−29; Norwegian mathematician. ○ *flooded walled plain (116 km)*

ADAMS (68E, 32S) 1. John Couch Adams, 1819−92; English astronomer, co-discoverer of Neptune; his calculations made independently of Le Verrier. 2. Charles H. Adams, 1868−1951; American amateur astronomer. 3. Walter Sydney Adams, 1876−1956; American astronomer; Director of Mt. Wilson Observatory. ○ *crater (66 km)*

AUSTRALE, MARE Southern Sea, see map 76.

BARNARD (86E, 29S) Edward Emerson Barnard, 1857−1923; American astronomer; discovered the fifth satellite of Jupiter; photographed galaxies; 'Barnard's star' in Ophiuchus has greatest known proper motion of any star.
○ *walled plain (120 km)*

FRAUNHOFER (59E, 40S) Joseph von Fraunhofer, 1787−1826; German optician; invented the diffraction grating and observed 'Fraunhofer's lines' in the solar spectrum.
○ *crater (57 km)*

FURNERIUS (60E, 36S) Georges Furner, about 1643; French Jesuit; Professor of Mathematics in Paris.
○ *prominent walled plain (125 km)*

GUM (89E, 40S) C. Gum, 1924−60; Australian astronomer.
○ *flooded, shallow crater (51 km)*

HAMILTON (43S, 84E) Sir William Rowan Hamilton, 1805−1865; Irish mathematician. ○ *regular, deep crater (57 km)*

MARINUS (76E, 39S) Marinus of Tyre, 2nd century AD; eminent geographer; first to point out that Asia and Africa might be larger than Europe and that the Roman Empire did not embrace the whole world. ○ *crater (58 km)*

OKEN (76E, 44S) Lorenz Oken (Okenfuss), 1779−1851; German biologist and physiologist. ○ *flooded crater (72 km)*

REICHENBACH (48E, 30S) Georg von Reichenbach, 1772−1826; German maker of geodesic and astronomical apparatus.
○ *crater (71 km)*

STEVINUS (54E, 32S) Simon Stevin, 1548−1620; Belgian mathematician, optician, soldier and engineer.
○ *prominent crater (74 km), central peak*

FURNERIUS

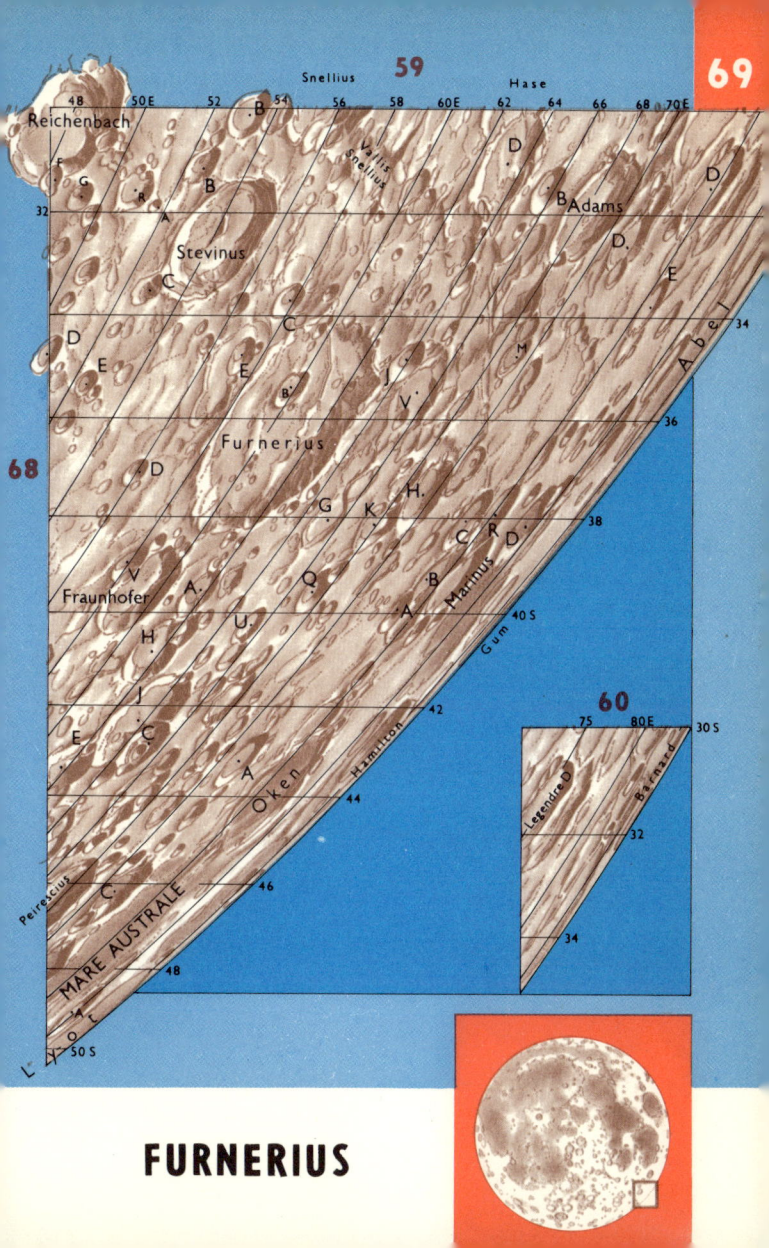

70 PHOCYLIDES

This area adjacent to the south-western limb of the Moon includes an exceptionally interesting crater, Wargentin, which is totally filled so that its floor resembles a plateau. The crater Pingré is situated on the inner edge of the wall belonging to the lunar basin, 300 km in diameter.

NASMYTH (56W, 50S) James Nasmyth, 1808–90; Scottish engineer; invented the steam hammer; selenographer; made models of the lunar surface. ○ *flooded crater (77 km)*

NÖGGERATH (46W, 49S) Jacob Nöggerath, 1788–1877; German geologist and mineralogist.
○ *crater with a flooded floor (31 km)*

PHOCYLIDES (57W, 53S) Johannes Phocylides Holwarda (Jan Fokker), 1618–51; Dutch astronomer; believed that the stars had their own motions.
○ *prominent walled plain with a flooded floor (114 km)*

PINGRÉ (74W, 59S) Alexandre Guy Pingré, 1711–96; French theologian and astronomer; author of 'Cometographie' (notes on comets; includes Uranus as it had just been discovered by Herschel and was still considered to be a comet).
○ *crater (85 km); before 1970 this crater was marked as Pingré A*

WARGENTIN (60W, 50S) Pehr Vilhelm Wargentin, 1717–83; Swedish astonomer; Director of Stockholm Observatory.
○ *the largest of very rare craters, which are filled to the top with dark material (lava); the crater's diameter is 84 km and its raised floor has numerous wrinkle ridges*

62

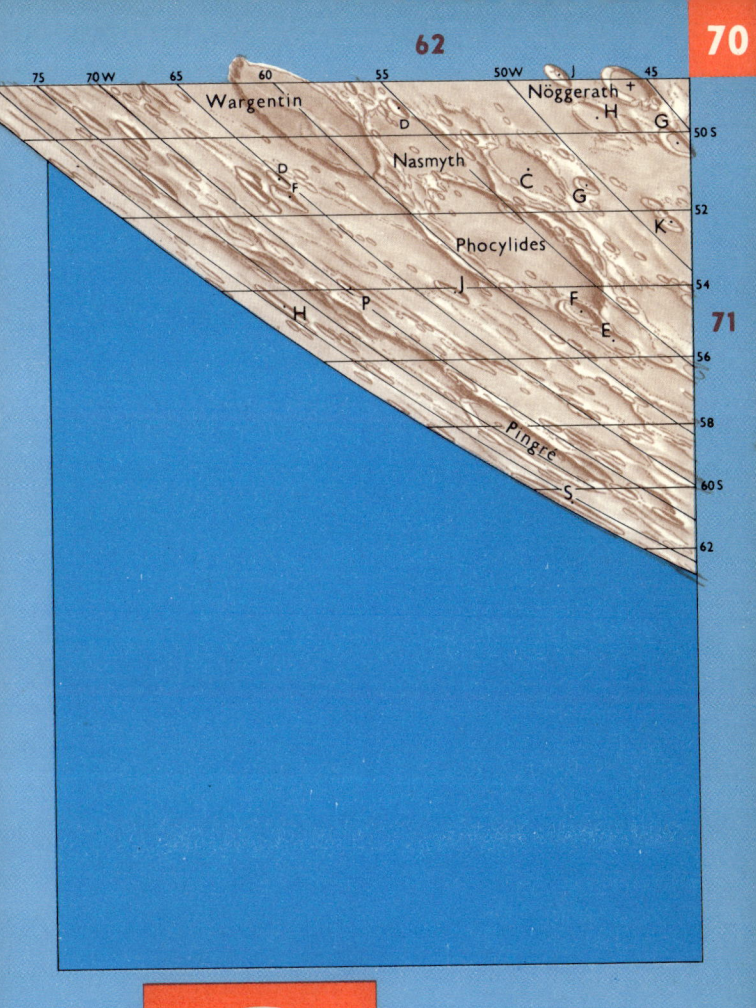

70

75 70W 65 60 55 50W 45

Wargentin   Nöggerath
D   H   G   50S
Nasmyth
F   C   G
P   K   52
Phocylides
J   F   54
H   P   E
71
56

Pingré   58

S   60S

62

PHOCYLIDES

71 SCHILLER

This south-western limb of the Moon contains two lunar basins; one of them is identical with crater Bailly and the second, unnamed basin, 350 m in diameter, is situated between the craters Schiller, Zuchius and Phocylides (see map 70).

BAILLY (60W, 67S) Jean Sylvain Bailly, 1736—93; French astronomer and politician.
○ *vast walled plain, identical to the lunar basin with outer wall 303 km in diameter*

BAYER (35W, 52S) Johann Bayer, 1572—1625; German; prepared the stellar atlas 'Uranometria', in which he introduced Greek letters to designate stars. ○ *crater (47 km)*

BETTINUS (45W, 64S) Mario Bettini, 1582—1657; Italian philosopher and mathematician. ○ *crater (71 km)*

HAUSEN (89 W, 65 S) Christian A. Hausen, 1693—1743; German astronomer, mathematician and natural philosopher.
○ *circular mountain range with central peaks (180 km); it is visible only during favourable librations*

KIRCHER (45W, 67S) Athanasius Kircher, 1601—80; German mathematician and professor of Oriental languages.
○ *crater with a flooded floor (72 km)*

ROST (34W, 56S) Leonhardt Rost, 1688—1727; German amateur astronomer and popularizer of astronomy.
○ *crater with a flooded floor (49 km)*

SCHILLER (40W, 52S) Julius Schiller, about 1627; German monk, author of 'Coelum Stellarum Christianum', Augsburg, 1627 (Christian Atlas of the Sky), in which the old traditional constellations are replaced by biblical characters and objects (it did not become popular).
○ *considerably elongated crater (179 × 71 km)*

SEGNER (48W, 59S) Johann A. von Segner, 1704—77; German natural philosopher; worked on the geometry of solar and lunar eclipses. ○ *shallow crater (67 km) with an undulating floor*

WEIGEL (39W, 58S) Erhard Weigel, 1625—99; German mathematician and astronomer; in his 'Astronomia Spherica' suggested replacing the traditional figures of the constellations by symbols of various countries. ('Coelum Heraldicum').
○ *crater (36 km)*

ZUCCHIUS (51W, 62S) Niccolo Zucchi, 1586—1670; Italian mathematician and astronomer, one of the first observers of the belts on Jupiter. ○ *crater (64 km)*

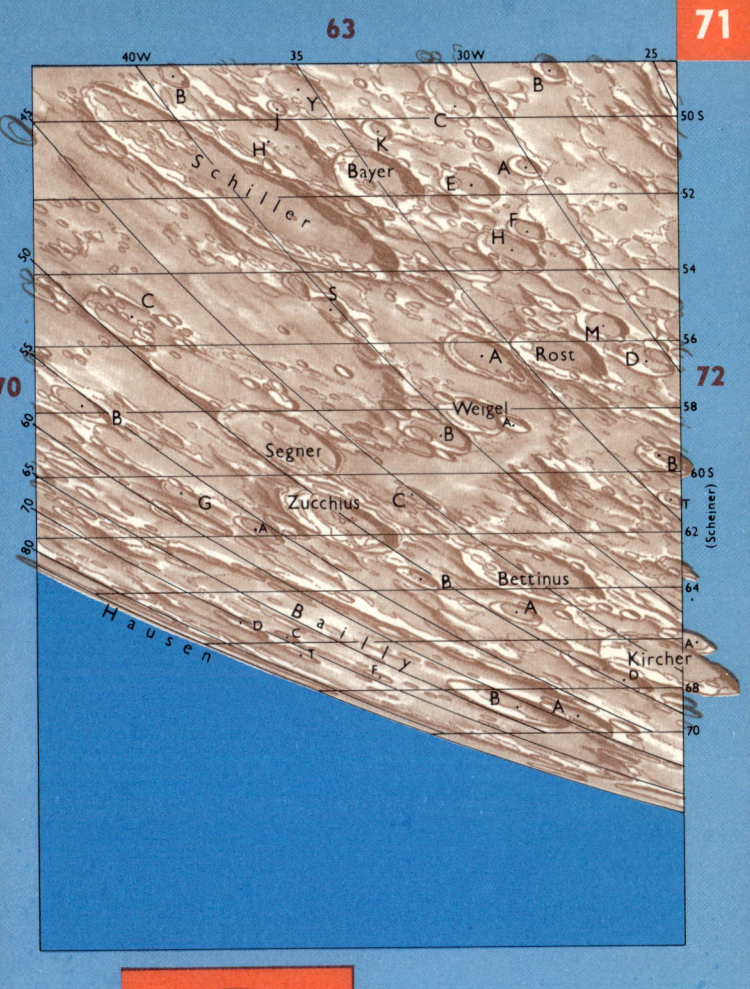

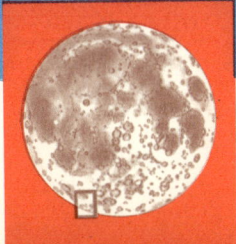

SCHILLER

72 CLAVIUS

The marginal area of the Moon adjacent to the south pole, which is densely covered by craters and large oval plains. The terrain is mountainous, especially close to the edge, and deep shadows make the observation and the mapping of this area very difficult; the same applies to the marginal areas on maps 73 and 74.

BLANCANUS (22W, 64S) Giuseppe Biancani, 1566−1624; Italian mathematician, geographer and astronomer.
○ *crater (105 km)*

CASATUS (30W, 73S) Paolo Casati, 1617−1707; Italian theologian and mathematician. ○ *flooded crater (111 km)*

CLAVIUS (14W, 58S) Christopher Klau, 1537−1612; German mathematician and astronomer, called 'Euclid of XVIth century'. ○ *one of the best known walled plains (225 km)*

DRYGALSKI (80W, 80S) Erich D. von Drygalski, 1865−1949; German geographer, geophysicist, polar explorer.
○ *circular mountain range (176 km), partly visible only during favourable libration*

KLAPROTH (26W, 70S) Martin Heinrich Klaproth, 1743−1817; German chemist and mineralogist.
○ *flooded walled plain (119 km)*

LE GENTIL (76W, 74S) Guillaume J. H. J. B. Le Gentil, 1725−92; French astronomer.
○ *greatly eroded crater (113 km)*

LONGOMONTANUS (22W, 50S) Christian Severin Longomontanus, 1562−1647; Danish astronomer, assistant to Tycho Brahe. ○ *walled plain (145 km)*

PORTER (10W, 56S) Russell W. Porter, 1871−1949; American, maker of telescopes.
○ *crater (52 km), until 1970 known as Clavius B*

RUTHERFURD (12W, 61S) Lewis Morris Rutherfurd, 1816−92; American astronomer; photography of the Sun and the Moon. ○ *crater (48 × 54 km)*

SCHEINER (28W, 60S) Christopher Scheiner, 1575−1650; German mathematician and astronomer; made first systematic observations of the Sun. ○ *crater (110 km)*

WILSON (42W, 69S) 1. Alexander Wilson, 1714−86; Scottish astronomer, friend of William Herschel. 2. Charles Thomson Rees Wilson, 1869−1959; Scottish physicist; noted for his invention of cloud chamber. 3. Ralph E. Wilson, 1866−1960; American astronomer at Mt. Wilson Observatory.
○ *considerably eroded crater (70 km)*

CLAVIUS

73 MORETUS

The South Pole lies in the centre of the bottom edge of the map. Observation of the area close to the pole is very difficult and some parts are constantly hidden behind the hills and crater walls.

AMUNDSEN (86E, 85S) Roald E. Amundsen, 1872−1928; famous Norwegian polar explorer; first to reach the South Pole in 1911; flew across the North Pole in 1928.

○ *crater (100 km)*
CABEUS (36W, 85S) Niccolo Cabeo, 1586−1650; Italian mathematician, philosopher. ○ *crater (about 95 km)*
CLAVIUS, see map 72.
CURTIUS (4E, 67S) Albert Curtz, 1600−71; German astronomer; published T. Brahe's observations. ○ *crater (95 km)*
CYSATUS (16W, 66S) Jean-Baptiste Cysat, 1588−1657; Swiss mathematician and astronomer. ○ *crater (49 km)*
DELUC (3W, 55S) Jean André Deluc, 1727−1817; Swiss geologist and physicist. ○ *crater (47 km)*
GRUEMBERGER (10W, 67S) Christoph Grienberger, 1561−1636; Austrian mathematician, astronomer ○ *crater (94 km)*
HERACLITUS (6E, 49S) About 540−480 BC; Greek philosopher of Ephesus. ○ *considerably disintegrated crater (90 km) with a central mountain range*
LILIUS (6E, 54S) Luigi Giglio, died in 1576; Italian natural philosopher; suggested reform of the Julian calendar.
○ *crater (61 km) with a central mountain*
MAGINUS (6W, 50S) Giovanni A. Magini, 1555−1617; Italian mathematician, astronomer. ○ *vast walled plain (163 km)*
MALAPERT (13E, 85S) Charles Malapert, 1581−1630; Flemish mathematician, poet and astronomer.
○ *irregular crater (about 55 km) close to the pole*
MORETUS (6W, 71S) Théodore Moretus, 1602−67; Flemish mathematician. ○ *prominent crater (114 km)*
NEWTON (17W, 77S) Isaac Newton, 1643−1727; English natural philosopher, law of gravitation. ○ *crater (64 km)*
PENTLAND (12E, 65S) Joseph B. Pentland, 1797−1873; Irish politician, explorer. ○ *crater (56 km)*
SHORT (7W, 75S) James Short, 1710−68; Scottish mathematician and optician. ○ *crater (71 km)*
SIMPELIUS (15E, 73S) Hugh Sempill (Sempilius), 1596−1654; Scottish mathematician. ○ *crater (70 km)*
ZACH (5E, 61S) Franz Xaver, Freiherr von Zach, 1754−1832; Hungarian astronomer, born in Bratislava. ○ *crater (71 km)*

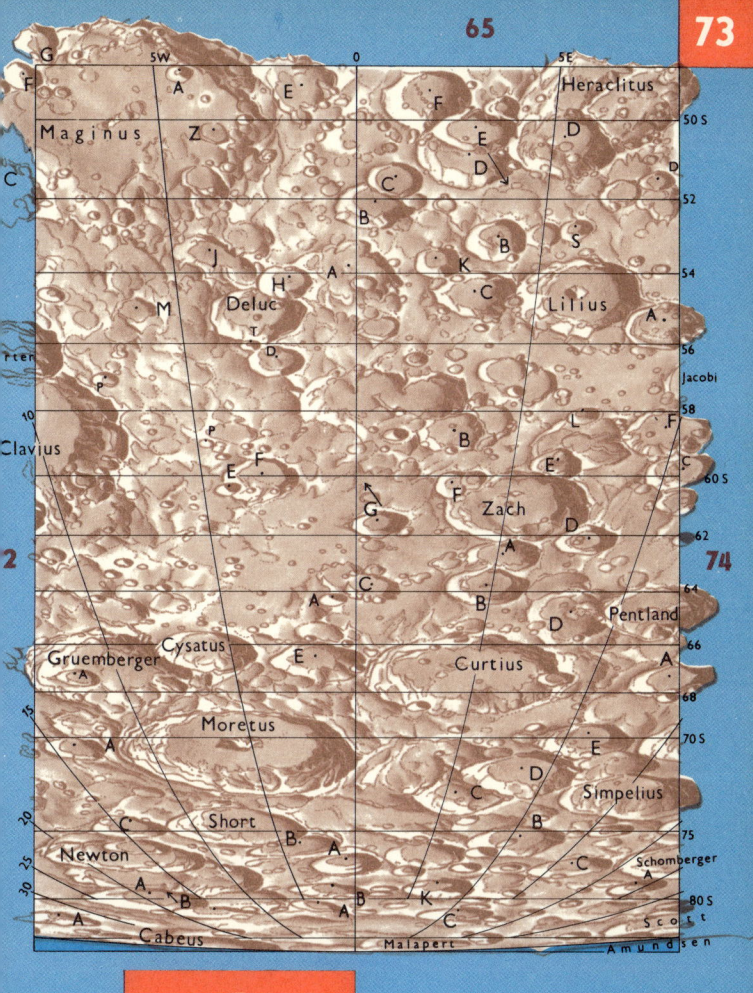

MORETUS

74 MANZINUS

The limb of the Moon east of the south polar region. The crater field in this area is very dense and chaotic. The pair of craters Mutus and Manzinus make the orientation easier.

ASCLEPI (26E, 55S) Giuseppe Asclepi, 1706–76; Italian Jesuit, astronomer and physicist. ○ *crater (42 km)*

BACO (19E, 51S) Roger Bacon, 1214–94; English philosopher, defender of the observational and experimental method against belief in authority. ○ *prominent crater (70 km)*

BOGUSLAWSKY (43E, 73S) Palon H. Ludwig von Boguslawsky, 1789–1851; German astronomer; discovered a comet in April 1835. ○ *crater with a flooded floor (97 km)*

BOUSSINGAULT, see map 75.

CUVIER (10E, 50S) Georges Cuvier, 1769–1832; French naturalist and palaeontologist.
○ *crater with a flooded floor (75 km)*

DEMONAX (59E, 78S) 2nd century BC; Greek philosopher, Cypriot by origin. ○ *crater (114 km)*

HALE, see map 75.

IDELER (22E, 49S) Christian Ludwig Ideler, 1766–1846; German chronologist. ○ *crater (39 km)*

JACOBI (11E, 57S) Karl Gustav Jacob Jacobi, 1804–51; German mathematician and philosopher; invented the Jacobian.
○ *crater with a flooded floor (68 km)*

KINAU (15E, 61S) C. A. Kinau, about 1850; German botanist and selenographer. ○ *crater (42 km), central peak*

MANZINUS (27E, 68S) Carlo A. Manzini, 1599–1677; Italian philosopher and astronomer.
○ *crater with a flooded floor (98 km)*

MUTUS (30E, 64S) Vincente Mut (or Muth), died in 1673; Spanish astronomer and navigator. ○ *crater (78 km)*

SCHOMBERGER (25E, 77S) Georg Schoenberger, 1597–1645; German mathematician and astronomer; believed that sunspots were satellites of the Sun ('stellae solares').
○ *crater (85 km)*

SCOTT (45E, 82S) Robert Falcon Scott, 1868–1912; English Antarctic explorer; second to reach the South Pole.
○ *crater with a disintegrated wall (108 km)*

TANNERUS (22E, 56S) Adam Tanner, 1572–1632; German mathematician and theologian.
○ *crater with a sharp rim (29 km)*

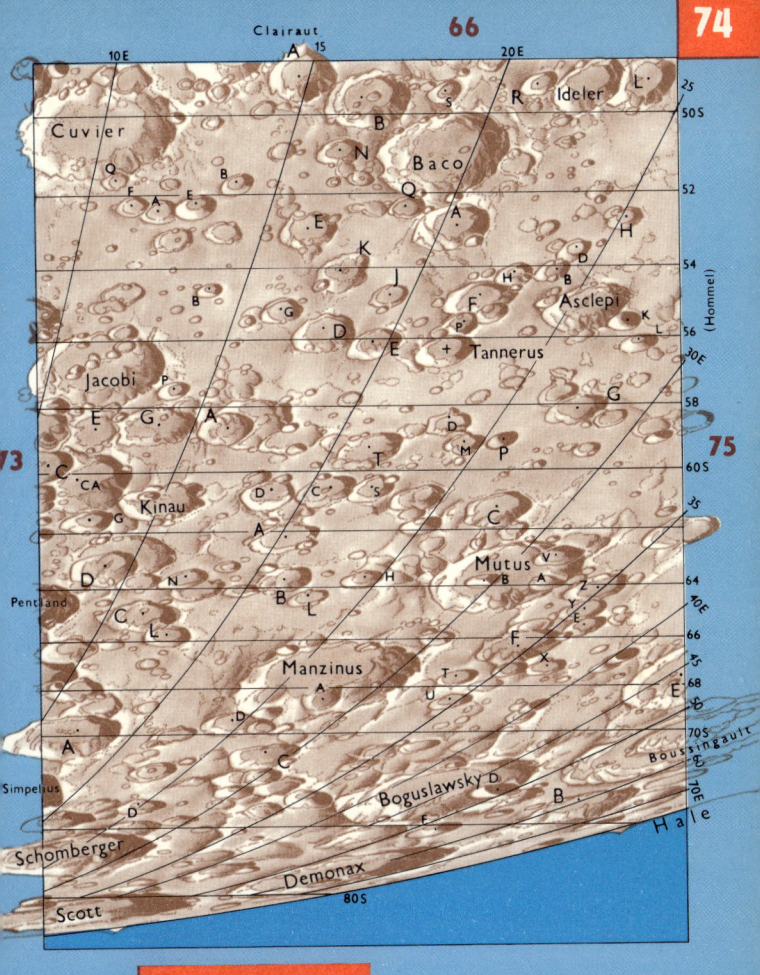

MANZINUS

75 HAGECIUS

A part of the south-eastern limb of the Moon with a group of large craters around Hagecius. Important for orientation is the crater Hommel, which together with smaller craters on its wall forms a characteristic group.

BIELA (51E, 55S) Wilhelm von Biela, 1782—1856; Austrian soldier and astronomer, Czech by origin; in 1826 discovered 'Biela's comet'. ○ *crater (76 km)*

BOUSSINGAULT (55E, 70S) Jean Baptiste Joseph Dieudonné Boussingault, 1802—87; French agricultural chemist and botanist. ○ *crater (131 km); inside the formation is the large crater Boussingault A, so that the whole feature resembles a crater with a double wall*

GILL (75E, 63S) Sir David Gill, 1843—1914; English astronomer. ○ *crater (64 km)*

HAGECIUS (47E, 60S) Thaddaeus Hayek (Tadeas Hajek of Hajek), 1525—1600; Czech natural historian, mathematician and astronomer; Tycho Brahe was invited to Prague at his advice. ○ *crater (76 km)*

HALE (90E, 74S) 1. George Ellery Hale, 1868—1938; American astronomer, Director of Mt. Wilson Observatory. 2. William Hale, 1797—1870; English scientist in the field of rocket technology. ○ *crater (80 km) stretching over to the far side*

HELMHOLTZ (64E, 68S) Hermann Ludwig Ferdinand von Helmholtz, 1821—1894; German physiologist and physicist. ○ *crater (95 km)*

HOMMEL (33E, 55S) Johann Hommel, 1518—62; German mathematician and astronomer, teacher of Tycho Brahe. ○ *crater (125 km)*

NEARCH (39E, 58S) Nearchus, about 325 BC; commander, officer of Alexander the Great. ○ *crater (75 km)*

NEUMAYER (71E, 71S) Georg B. von Neumayer, 1826—1909; German meteorologist. ○ *crater (76 km)*

PITISCUS (31E, 50S) Bartholomaeus Pitiscus, 1561—1613; German mathematician. ○ *prominent crater (82 km)*

ROSENBERGER (43E, 55S) Otto A. Rosenberger, 1800—90; German mathematician and astronomer. ○ *crater (96 km)*

VLACQ (39E, 53S) Adriaan Vlacq, died about 1660; Dutch bookseller and mathematician; in 1628 he published logarithmic tables calculated to 10 places. ○ *prominent crater (89 km)*

WEXLER (90E, 69S) Harry Wexler, 1911—62; American meteorologist; programme of weather satellites.
○ *crater (50 km) stretching over to the far side of the Moon*

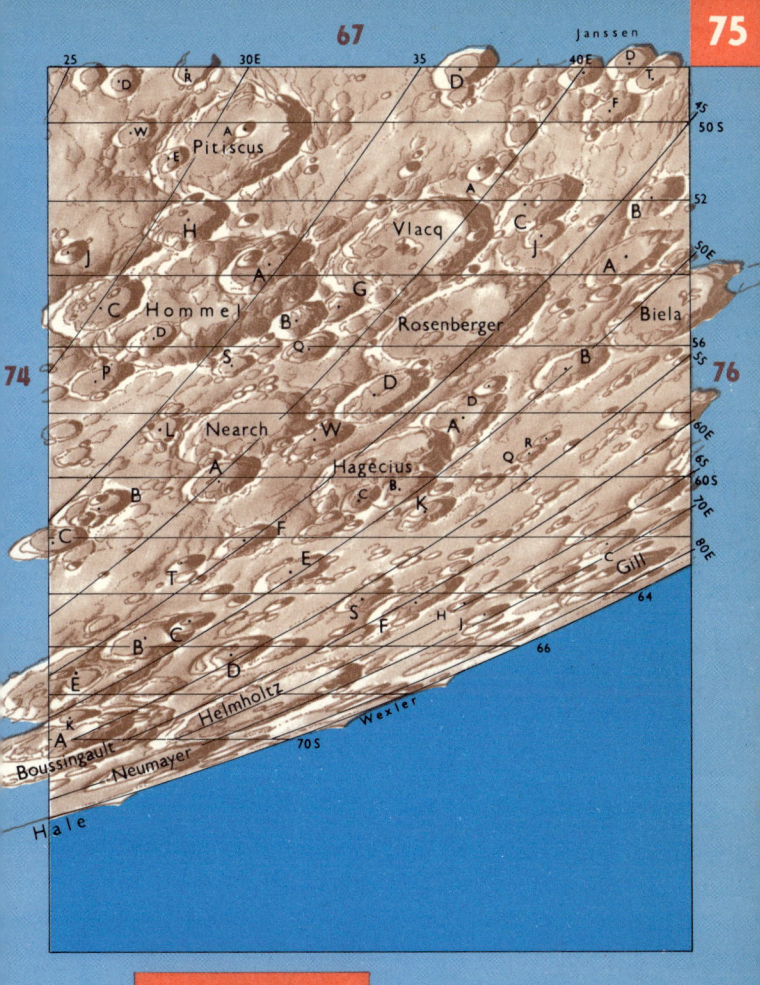

HAGECIUS

76 WATT

The south-eastern limb of the Moon. The numerous dark patches on its edge form Mare Australe. The contours of M. Australe, as determined from photographs made by Lunar Orbiters, are circular and the diameter is about 900 km, so that it is from an evolutionary point evidently a very old lunar basin.

AUSTRALE, MARE Southern Sea. Irregular lunar mare, which stretches to the far side of the Moon. Its surface area is 148 000 sq km and in many places is covered by craters and light areas, which resemble small islands.

BRISBANE (68E, 49S) Sir Thomas Makdougall Brisbane, 1770–1860; Scottish by origin, soldier, politician, astronomer.
○ *crater (45 km)*

HANNO (71E, 56S) About 500 BC; Carthaginian navigator, who sailed round the African coast. ○ *crater (56 km)*

LYOT (85E, 51S) Bernard Ferdinand Lyot, 1897–1952; French astronomer; invented the solar coronograph and monochromatic polarizing filter ('Lyot's filter').
○ *flooded walled plain (141 km) with decayed border and a dark, circular floor*

PETROV (88E, 61S) Evgenii S. Petrov, 1900–42; Soviet scientist in the field of rocket technology.
○ *flooded crater (50 km)*

PONTÉCOULANT (66E, 59S) Philippe G. Doulcet, Comte de Pontécoulant, 1795–1874; French mathematician and astronomer; forecast the return of Halley's comet in 1835 to within 3 days. ○ *prominent crater (91 km)*

STEINHEIL (46E, 49S) Karl August von Steinheil, 1801–70; German mathematician, physicist, optician and astronomer.
○ *prominent crater with a flooded floor (67 km)*

WATT (49E, 50S) James Watt, 1736–1819; Scottish engineer, invented the improved steam engine; Watt's governor used for controlling mechanical drive of the telescope.
○ *crater (66 km); the western edge is formed by a wall of the neighbouring crater Steinheil, with which Watt forms a characteristic pair*

68 76

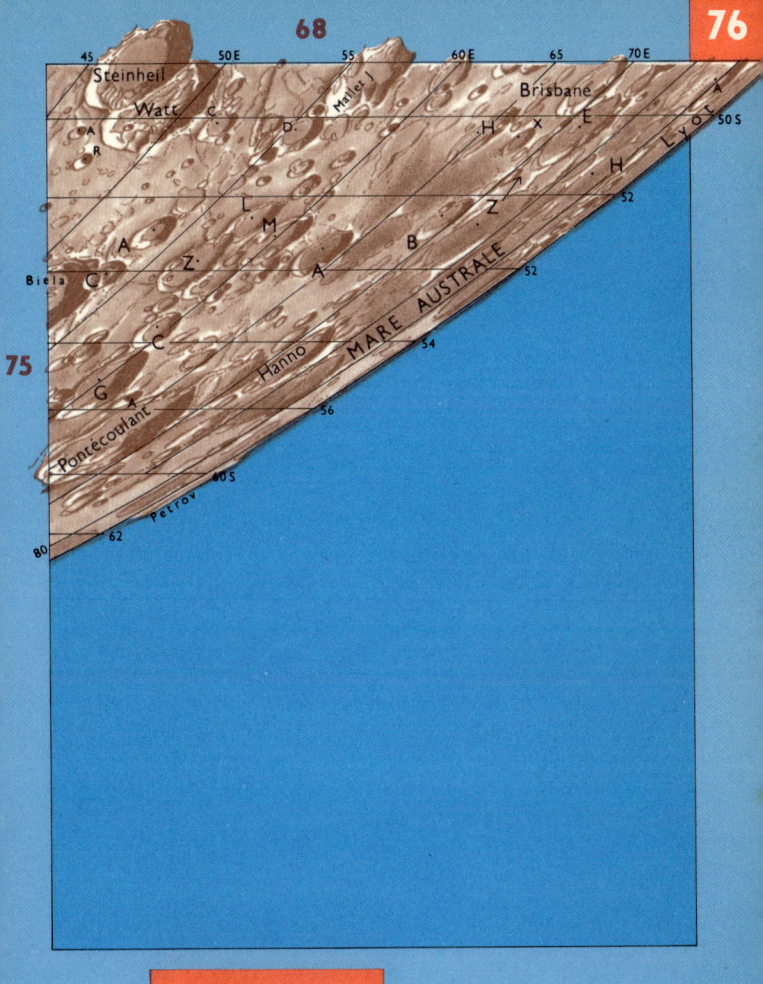

45 50 E 55 60 E 65 70 E
Steinheil
Watt c Mallet Brisbane
A D H x E 50 S
R H
L M Z 52
A Z A B
Biela C MARE AUSTRALE 52
C Hanno
75
G 54
Pontécoulant 56
80 62 Petrov 60 S

WATT

MARS

A. The Equatorial Region of Mars between Areographic Longitudes 315° and 45°

The centre of the map is intersected by the prime meridian. This and other maps recording the equatorial regions (A, B, C and D) are bordered in the north and the south by latitudes 60N and 60S; these latitudes are adjacent to the winter polar regions and are usually covered by the polar caps and white clouds.

In the north this area is penetrated by a very dark and clearly visible Mare Acidalium, which is separated from the oblong 'lake' Niliacus Lacus by the light band of Achillis Pons. The wide canal Deuteronilus stretches east as far as the lake Ismenius Lacus, whose size and shape fluctuate appreciably.

The very prominent dark strip of Sinus Sabaeus stretches along the equator and in the east twists and changes into the bow-like Mare Serpentis and close to the centre of the map links up with the well-known Sinus Meridiani, which marks the beginning of the areographic coordinates. Larger telescopes clearly reveal the two characteristic promontories extending north from Sinus Meridiani.

The light isthmus Aram separates Sinus Meridiani from the dark triangular bay Margaritifer Sinus, which is clearly visible even through small telescopes, and which in the south merges with Mare Erythraeum. Further south is the red oval-shaped continent Argyre I. It sometimes turns paler or almost white, so that a beginner might confuse it with the polar cap. In reality Argyre I is a basin about 1 000 km in diameter and has a smooth floor.

Sinus Sabaeus lies south of the light island Deucalionis Regio, which is separated from the continent Noachis by a grey narrow strip, Pandorae Fretum; this is a typical example of a formation which is subject to sudden seasonal changes (see pp. 49−50).

The relief of the whole area is generally uniform, and consists mainly of crater fields. An exception is the smooth surface of Argyre I and another similar craterless but slightly undulating area in the region of Niliacus Lacus and Deuteronilus. The dark surface of Mare Acidalium is also smooth. A large canyon originates in the western edge of Margaritifer Sinus, and continues further westward across Aurorae Sinus.

A

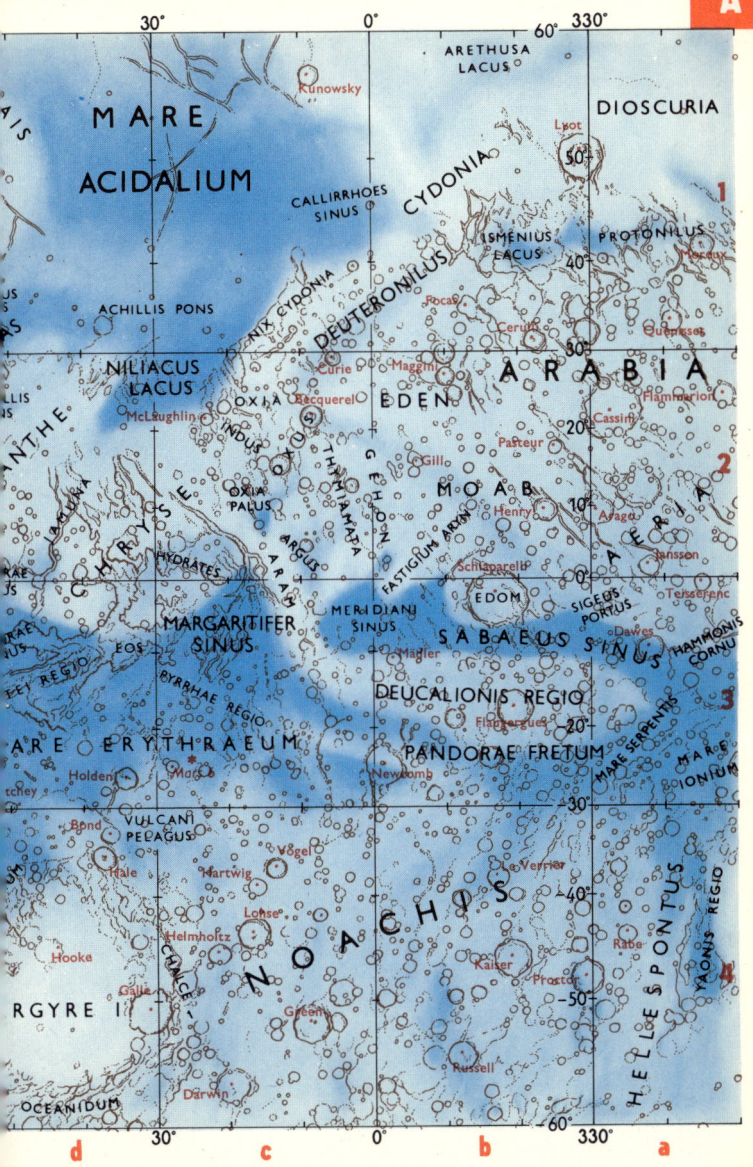

MARE ACIDALIUM

DIOSCURIA

- Kunowsky
- ARETHUSA LACUS
- Lyot
- CALLIRRHOES SINUS
- CYDONIA
- ISMENIUS LACUS
- PROTONILUS
- Mocyr
- ACHILLIS PONS
- PHOENIX CYDONIA
- DEUTERONILUS
- Focas
- Cerulli
- Quenisset

ARABIA

- NILIACUS LACUS
- OXIA
- Curie
- Maggini
- EDEN
- Becquerel
- McLaughlin
- Flammarion
- INDUS
- OXUS
- THYMIAMATA
- Pasteur
- Cassini
- CHRYSE
- GEHON
- Gill
- MOAB
- Henry
- Arago
- Janssen
- OXIA PALUS
- HYDRATES
- ARAXES
- FASTIGIUM ARYN
- Schiaparelli
- AERIA
- ARGUS
- MERIDIANI SINUS
- EDOM
- SIGEUS PORTUS
- Dawes
- Teisserenc
- MARGARITIFER SINUS
- SABAEUS SINUS
- HAMMONIS CORNU
- EOS
- Mädler
- DEUCALIONIS REGIO
- BYRRHAE REGIO
- Flaugergues
- MARE SERPENTIS
- REGIO
- ARE ERYTHRAEUM
- PANDORAE FRETUM
- Newcomb
- MARE IONIUM
- Holden
- Mac 6
- Bond
- VULCANI PELAGUS
- Vögel
- Hale
- Hartwig
- Le Verrier
- Rabe
- Lohse
- NOACHIS
- YAONIS REGIO
- Hooke
- Helmholtz
- Kaiser
- Proctor
- Galle
- HELLESPONTUS
- RGYRE I
- Green
- Russell
- OCEANIDUM
- Darwin

d 30° c 0° b 330° a

B. The Equatorial Region of Mars between Areographic Longitudes 45° and 135°

The centre of the map is intersected by the 90° meridian, which runs across Solis Lacus. The area west of the 70° meridian in the northern hemisphere is regarded as a 'difficult region', because it contains minute details which are difficult to observe. Small telescopes clearly show Nilokeras, which projects south-west of Mare Acidalium, and a weaker, indefinite patch, Lunae Palus. Nix Olympica, which belongs to the most interesting formations in the area north of the equator, is sometimes visible as a very bright, circular white patch, but on other occasions it is not visible at all.

A veritable wealth of detail, visible even through small telescopes, is provided by the region south of the equator. First of all there is Solis Lacus, surrounded by the light surface of Sinai, Thaumasia, Claritas and Syria. This complex is encircled by grey patches, canals (Coprates, Bosporus) and small lakes (Lacus Melas, Tithonius Lacus, Noctis Lacus, Phoenicis Lacus). The whole region resembles the shape of an eye, and is often called the 'Eye of Mars' by astronomers. It is also subject to long-term periodic changes.

An interesting detail is the lake Juventae Fons, which at times is one of the darkest areas on Mars. South of the 'Eye of Mars' is the relatively monotonous grey surface of Aonius Sinus and a section of the vast Mare Australe.

The relief of this region is exceptionally interesting. By far the greatest proportion is flat, especially the area north of the equator. The peaks of giant volcanoes rise from the plain, the highest being located in Nix Olympica. Other volcanoes are found in the area of Pavonis Lacus and Ascraeus Lacus. The light patches and white clouds, which temporarily appear in this area, can be explained by the presence of high mountainous features, which cause the condensation of clouds and the formation of hoarfrost.

A most unusual feature is provided by a vast canyon, the dimensions of which are such that it is even partially visible through small terrestrial telescopes. The eastern, branched and widened part of the canyon is known as Aurorae Sinus. To the west it continues as the canal Coprates, then it widens again and branches out in Melas Lacus, and finally terminates as a row of patches on the southern edge of Tithonius Lacus; further to the west it continues as a maze of deep clefts in Noctis Lacus. This is one of the few instances that reveal an evident connection between differences in albedo and the morphology of the terrain.

B

Map Regions and Features

120° — **90°** — **60°**

- Mankovitch
- TANAIS
- 50° TEMPE
- ARCADIA
- ALBA
- Barabashov
- 1
- LYCUS
- 40°
- IDAEUS FONS
- CYANE
- 30°
- NILOKERAS
- Perepelkin
- ACHILLIS FONS
- IX OMPICA
- 20° Fessenkov
- LUNAE LACUS
- XANTHE
- THARSIS
- 2
- ASCRAEUS LACUS
- 10° CANDOR
- PAVONIS L.
- GANGES
- HYDRAE PALUS
- 0° TITHONIUS LACUS
- JUVENTAE FONS
- ARSIA SILVA
- COPRATES
- OPHIR
- BAETIS
- −10° PHOENICIS LACUS
- Oudemans
- MELAS LACUS
- AURORAE SINUS
- SYRIA
- COPRATES
- ARAXES
- SINAI
- PROTEI REGIO
- 3
- −20° DAEDALIA
- Lassell
- MARE
- ITUS
- SOLIS LACUS
- NECTAR
- Holden
- BATHYS
- THAUMASIA
- Ritchey
- −30° SIRENUM SINUS
- Bond
- Pickering
- BOSPOROS
- NEREIDUM FRETUM
- ICARIA
- −40°
- AONIUS SINUS
- PONTICA DEPRESSIO
- Hooke
- 4
- CHRYSOKERAS
- Slipher
- −50°
- Porter
- Lowell
- OGYGIS REGIO
- Halley
- ARGYRE
- Hussey
- Brashear
- Clark
- Lamont
- Ross
- Coblentz
- MARE OCEANIDUM

−60°

120° — **90°** — **60°**

d — c — b — a

C. The Equatorial Region of Mars between Areographic Longitudes 135° and 225°

The 180° meridian runs through the centre of this map. In the northern hemisphere, the 'difficult region' in map B continues as far as areographic longitude 185°. This area mostly contains only indefinite and obscure grey patches, which in the past were considered to be part of the illusory canals. A very striking feature is the lake Trivium Charontis, from which the dark, wide canal Cerberus runs southwards, while to north-north-west is the less prominent Styx. Both canals lie along the edge of the continent Elysium, which is one of the lightest areas on Mars.

A great many details can be seen through the telescope in the zone of dark seas on the southern hemisphere. There is in particular the prominent dark arch of Mare Sirenum, with three characteristic bays, Sirenum Sinus, Gorgonum Sinus and Titanum Sinus in the north. Mare Cimmerium is linked with Mare Sirenum by a large arch with a series of projections — Laestrygonum Sinus, Cyclopum Sinus, Gomer Sinus and Tritonis Sinus (see also map D).

Below latitude 40S the seas are replaced by a zone of light, red-coloured continents, namely Phaethontis, Electris and Eridania. The landing section of the probe Mars 3 made its soft landing on the dividing line between Phaethontis and Electris. The continent Electris continues south of latitude 55S into the dark patch of Mare Chronium.

The relief of region C is relatively monotonous. North of areographic latitude 10S, the smooth terrain lacks any significant features. The continental region of Amazonis and Mesogaea is particularly smooth. The terrain in the area of Trivium Charontis is chaotic. The surface of Elysium is relatively flat, except for two isolated mountains and a network of wide clefts in the western part of Elysium.

An almost continuous field of shallow craters stretches south of latitude 10S as far as the southern edge of the map. It is interesting that such a sharp division between sea and continent, as is featured here between Mare Sirenium and the light region of Memnomia, has no counterpart in the surface relief — the continuous crater field runs from the dark into the light area without any interruption. On the other hand, the northern edge of Laestrygonium Sinus and Gomer Sinus are approximately identical with the dividing line between the crater field in the southern hemisphere and the vast, smooth continental surfaces of Zephyria and Aeolis.

C

Map labels

210° · **180°** · **150°**

- ONIUS CUS
- ANIAN
- Mie
- CEBRENIA
- Stokes
- PROPONTIS II
- HERCULIS PONS
- DIACRIA
- CASTORIUS LACUS
- Milankovitch
- ARCADIA
- 50°
- Tyndall
- EUXINUS LACUS
- HECATES LACUS
- PROPONTIS I
- 40°
- CHAOS
- HYBLAEUS
- Adams
- 30°
- Lockyer
- STYX
- PHLEGRA
- AMAZONIS
- LYCUS
- ELYSIUM
- EREBUS
- 20°
- CYANE
- EUNOSTOS
- Eddie
- TRIVIUM CHARONTIS
- Pettit
- NIX OLYMPICA
- CERBERUS
- 10°
- MESOGAEA
- CYCLOPIA
- ZEPHYRIA
- 0°
- Nicholson
- NODUS GORDII
- ARS
- SINUS
- AEOLIS
- −10°
- TITANIUM SINUS
- MEMNONIA
- Gale
- Boeddicker
- GIGANTUM SINUS
- Burton
- AESTAGONIUM SINUS
- −20°
- Williams
- CIMMERIUM
- BASENA
- GORGONUM SINUS
- Molesworth
- MARE
- −30°
- Columbus
- Hartz
- ATLANTIS
- MARE CRENUM
- Pickering
- SIRENI SINUS
- Cruls
- SALAMONDER
- −40°
- Newton
- ERIDANIA
- ELECTRIS
- PHAETHONTIS
- Kepler
- Huggins
- Mars 2
- Tycho Brahe
- −50°
- Copernicus
- Campbell
- Liu Shin
- Hussey
- TIPHYS FRETUM
- Mendel
- PALINURI FRETUM
- Wright
- Clark

210° · **180°** · **150°**

d · c · b · a

1 · 2 · 3 · 4

D. The Equatorial Region of Mars between Areographic Longitudes 225° and 315°

The 270° meridian forms the vertical axis of this map. This part of Mars is probably the most famous on account of the very striking, dark triangular patch Syrtis Major. This formation can be observed even through telescopes with an aperture of 50 mm. Syrtis Major displays considerable seasonal changes. Only its western edge remains constant, while eastwards, towards Moeris Lacus, this formation regularly widens out in mid-summer as if to absorb the continent of Libya and lake Moeris Lacus. In spring and at the beginning of summer Syrtis Major is relatively narrow.

In the area of the canal Nepenthes, east of Syrtis Major, numerous irregular periodic changes of considerable extent have been observed. These changes have resulted in the appearance and subsequent complete disappearance of a vast 'sea'. The prominent dark regions of Casius and Nilosyrtis are north of Syrtis Major. The narrow canal Protonilus projects westward from the small round patch Coloe Palus.

Syrtis Major is separated by an indistinct isthmus, Oenotria, from the bay Deltonon Sinus, which is further connected with Sinus Sabaeus. South-east of Syrtis Major there stretches the wide, dark strip of Mare Tyrrhenum with its projection Syrtis Minor.

The circular light continent Hellas in the southern hemisphere is usually a very impressive formation as it undergoes distinct seasonal changes. As the 'dark wave' gathers momentum, the originally grey surface of Hellas starts to become brighter, until the whole area assumes a sharper outline and the small dark patch of Zea Lacus takes shape within. Sometimes Hellas is so light that it resembles the polar cap.

The relief of region D is varied. The continent Hellas is a vast basin with a smooth, featureless surface. The diameter of the basin is 1 400 km, its depth is over 6 000 m and its lowest point is about 32 km lower than the highest peaks on the planet. A vast crater field runs north from Hellas as far as the equator, where it terminates in the area Oenotria-Crocea — the western edge of Libya — the southern edge of Moeris Lacus. The relief seemingly follows the albedo, but the whole of the light area of Libya is strewn with craters as in the case of the neighbouring Mare Tyrrhenum.

Syrtis Major has a relatively smooth surface, which slowly rises toward the continent of Aeria, up to a height of about 6 000 m. The crater field Aeria ends along the western edge of Syrtis Major, and further to the east of Syrtis a dark, featureless surface stretches north to latitude 10N.

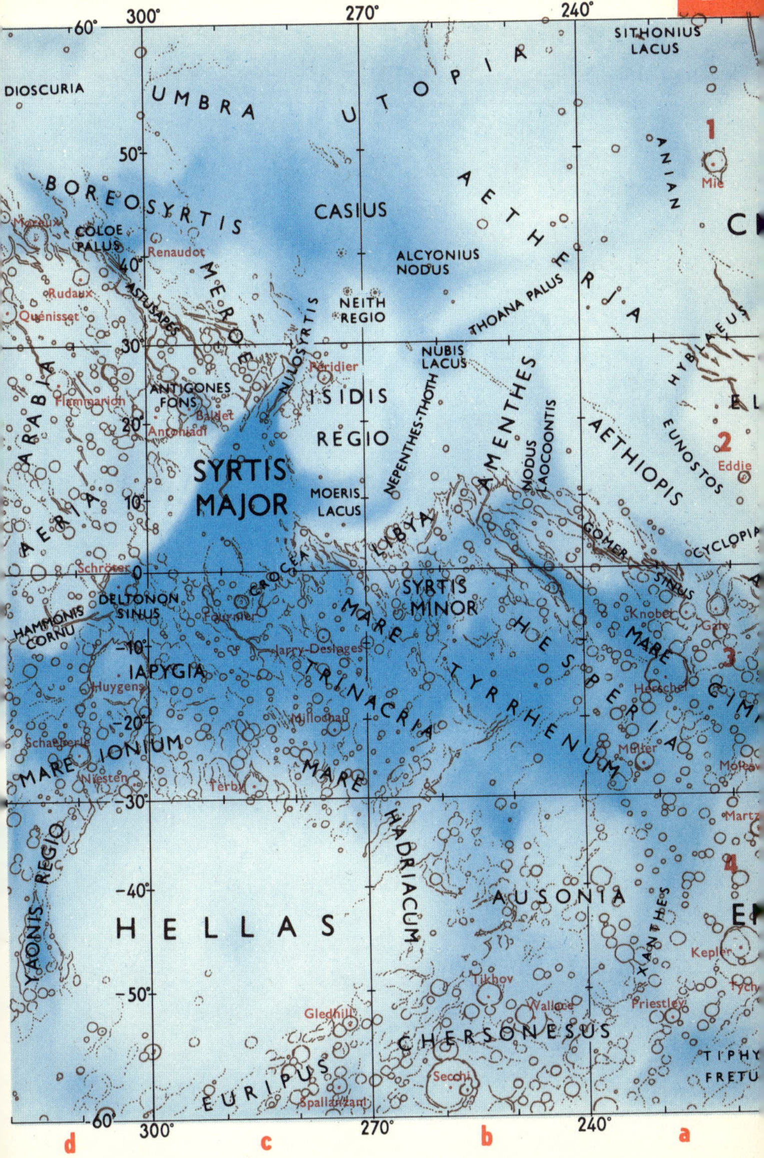

D

Map labels (Mars chart)

60° — 300° — 270° — 240°

- SITHONIUS LACUS
- DIOSCURIA
- UMBRA
- UTOPIA
- **50°**
- AETHERIA
- 1
- BOREOSYRTIS
- CASIUS
- ANIAN
- Mie
- **40°**
- Mazaeus
- COLOE PALUS
- Renaudot
- ALCYONIUS NODUS
- C
- Rudaux
- MEROE
- NEITH REGIO
- Quénisset
- THOANA PALUS
- HYBLAEUS
- **30°**
- Kéridier
- NUBIS LACUS
- ARABIA
- ANTIGONES FONS
- NILOSYRTIS
- ISIDIS REGIO
- AMENTHES
- AETHIOPIS
- EUNOSTOS
- 2
- Flammarion
- Baldet
- NEPENTHES-THOTH
- Eddie
- **20°**
- Antoniadi
- SYRTIS MAJOR
- MOERIS LACUS
- NODUS LAOCOONTIS
- AERIA
- **10°**
- Schröter
- CROCEA
- LIBYA
- GOPHER
- CYCLOPIA
- STELLA
- **0°**
- DELTOTON SINUS
- Fournier
- SYRTIS MINOR
- HAMPIONIS CORNU
- Knobel
- 3
- IAPYGIA
- Jarry-Desloges
- Hoechst
- Gale
- Huygens
- MARE TYRRHENUM
- MARE HESPERIA
- **-20°**
- Schaeberle
- Millochau
- MARE IONIUM
- MARE TRINACRIA
- Müller
- Mowt
- Niesten
- Terby
- **-30°**
- MARE HADRIACUM
- Martz
- **-40°**
- AUSONIA
- 4
- YAONIS REGIO
- XANTHES
- E
- **-50°**
- HELLAS
- Kepler
- Tikhov
- Wallace
- Lychy
- Gledhill
- Priestley
- CHERSONESUS
- TIPHY FRETU
- **-60°**
- EURIPUS
- Spallanzani
- Secchi

300° — d | 270° — c | 240° — b | a

E. The North Polar Region of Mars

The North Pole of Mars is at the centre of the map and its perimeter is given by areographic latitude 50N.

Prior to the advent of interplanetary probes, when astronomers were entirely dependent on telescopic observations, the area surrounding the North Pole was the least investigated region of Mars. The reason is that this area can be observed only during aphelic oppositions, when the North Pole is inclined to the Earth. However, Mars is then very distant from the Earth and therefore little detail is visible. Furthermore, these polar regions can be observed only in spring and summer in the northern hemisphere. In the autumn and in winter the extensive area surrounding the pole disappears as far as areographic latitude 45N (or even further) under an opaque cover of white polar mist, under which the polar cap starts to form. At the beginning of spring the cap starts to appear as a white, circular area, which reaches approximately latitude 65N. Exceptionally, the cap stretches as far as latitude 45N, but it is always smaller than the south polar cap.

During spring and summer the cap grows smaller, being bordered by a dark polar strip. As the cap shrinks, it gives way to light and dark areas, which are recorded here on the map. The most striking dark formation is Lacus Hyperboreus, which roughly extends from the pole in the direction of Mare Acidalium. This area, as far as latitude 60N, is very poor in details. However, a ring of dark patches composed of Mare Boreum, Scandia, Panchaia and Copais Palus starts south of this line.

The north polar cap never disappears completely, although during summer it shrivels to a small area about 6° in diameter. Its centre is not identical with the pole, but lies about 1° away on longitude 290°. When the cap decreases the small, white island Olympia regularly detaches itself from it and stretches along latitude 80N.

At the end of its programme, Mariner 9 succeeded in acquiring the first information about the physical relief of Mars in the vicinity of the North Pole. The local situation is very similar to that of the South Pole. The crater fields and plains with a small number of craters reach as far as latitude 80N. Closer to the pole is a smooth area with sedimentary layers. These layers reach a thickness of up to 100 m and their total height exceeds 1 km. The remains of the polar ice and hoarfrost run approximately parallel to winding low mountain ranges and to the edges of sedimentary layers.

E

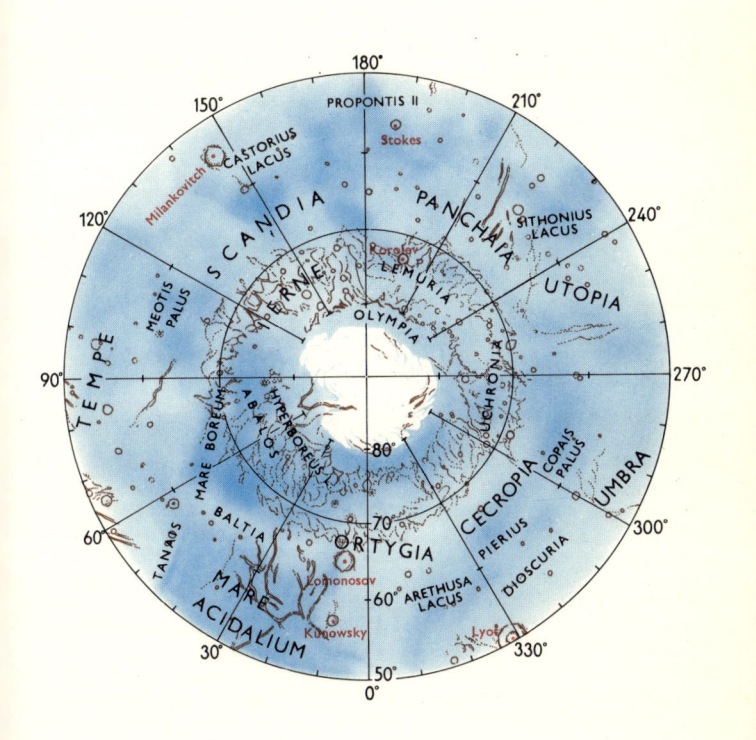

F. The South Polar Region of Mars

The South Pole of Mars is in the centre of the map and its perimeter is given by areographic latitude 50S. The south polar region is best visible during perihelic oppositions, when Mars is closest to the Earth. It is therefore better known than the north polar region, although even the south polar region does not reveal much more detail.

Undoubtedly the most interesting object for amateur observers is the polar cap. Its development is similar to that of the North Pole, which is described in map E. When the southern cap appears after dispersal of the winter layer of mist, it is usually 60° to 70° in diameter, so that it reaches latitude 60S to 55S. It is always larger than the northern cap, because winter is longer and colder in the southern hemisphere. The short yet hot summer makes the cap melt very quickly, so that its disappearance progresses daily. The shape of the cap also changes, and a part of it becomes separated from the rest of the cap. Even small telescopes clearly reveal the development of the dark 'cleft' (Rima Australis), which separates from the cap to form a narrow strip called Novus Mons.

The melting cap contracts towards the pole, and during summer only a small round patch about 5° in diameter remains. Its centre is not identical with the pole, being about 3° away from the south pole on areographic longitude 40°.

The south polar region is also subject to seasonal changes. For example, during the vernal equinox, a dark triangle protruding from the polar cap towards Hellespontus is clearly visible.

A circle of dark patches, namely Mare Chronium, Promethei Sinus, Depressio Hellespontica and Mare Australe, run along latitude 60S. When the polar cap recedes, light continents Thyle I and Thyle II appear in its place.

The relief of the south polar region is quite complex. The crater fields and smooth plains dotted with a small number of craters stretch towards latitude 75S to 80S. This level is also reached by the arch of mountain ranges which stretch from Mare Chronium across Promethei Sinus towards Mare Australe. Above latitude 80S, the relief resembles the area of the North Pole — the smooth surface of sedimentary layers is sporadically replaced by arching low mountain ranges.

F

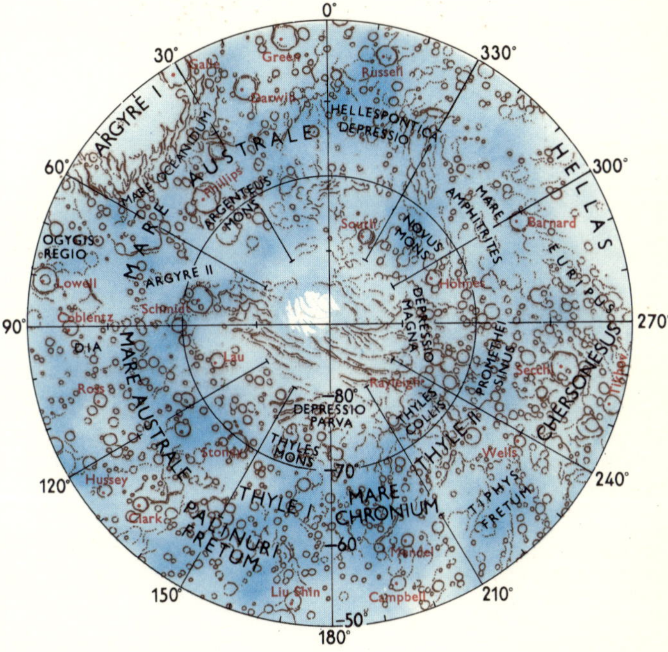
0°
30°
330°
60°
300°
90°
270°
120°
240°
150°
210°
180°
-80°
-70°
-60°
-50°
ARGYRE I
MARE AUSTRALE
MARE OCEANIDUM
HELLESPONTICA DEPRESSIO
HELLAS
MARE AMPHITRITES
ARGENTEUS MONS
NOVUS MONS
OGYGIS REGIO
ARGYRE II
EURIPUS
DEPRESSIO MAGNA
DIA
MARE AUSTRALE
PROMETHEI SINUS
CHERSONESUS
DEPRESSIO PARVA
THYLES COLLIS
THYLES MONS
THYLE II
THYLE I
MARE CHRONIUM
TIPHYS FRETUM
PALINURI FRETUM
Galle
Green
Russell
Darwin
Phillips
South
Barnard
Lowell
Holmes
Schmidt
Coblentz
Lau
Secchi
Ross
Rayleigh
Stoney
Wells
Hussey
Clark
Mendel
Liu Shin
Campbell

Comparative Table of Data on the Moon and Planets

	EARTH	MOON	MERCURY	VENUS	MARS
Distance from the Sun in *Mkm*	149.6	—	57.9	108.2	227.9
Sidereal period in *days*	365.256	27.322	87.969	224.7	687
Distance from the Earth *(Mkm)*					
minimum	—	0.356	79	39	55.5
maximum	—	0.406	218	260	400.0
Diameter *in km*	12 756	3 476	4 880	12 100	6 790
Earth = 1	1	0.27	0.38	0.95	0.53
Sidereal period of axial rotation	23h56m04s	27d07h43m	58d16h	243 days	24h37m23s
Volume (Earth = 1)	1	0.02	0.055	0.857	0.15
Mass (Earth = 1)	1	0.012	0.055	0.815	0.107
Mean density g/cu cm	5.52	3.34	5.5	5.25	3.94
Surface gravity (Earth = 1)	1	0.16	0.38	0.90	0.38
Escape velocity km/s	11.2	2.4	4.3	10.3	5.0

List of craters on the near side of the Moon, recently named by the I. A. U. in Sydney, 1973

Map number	Crater	Old name	Selenographic longitude	latitude	Diameter (km) Depth (m)	Personality
37	Abbot	Apollonius K	54.7 W	6.0 N	10	US solar physicist (1872 – 1973)
23	Banting	Linné E	16.4 W	26.5 N	6/1100	Canadian endocrinologist (1891 – 1941)

241

Map number	Crater	Old name	Selenographic longitude	latitude	Diameter (km) Depth (m)	Personality
23	Bowen	Manilius A	9.0 W	17.6 N	9/1100	US astronomer (1898–1973)
24	Brackett		23.6 W	18.0 N	8	US physicist (1896–1972)
36	Cajal	Jansen F	31.0 W	12.7 N	10/1800	Spanish histologist (1852–1934)
37	Cameron	Taruntius C	45.9 W	6.3 N	11	US astronomer (1925–72)
25	Carmichael	Macrobius A	40.3 W	19.6 N	20/3640	US psychologist (1898–1973)
25	Clerke	Littrow B	29.2 W	21.8 N	7/1430	US historian of astronomy (1842–1907)
26	Curtis		56.5 W	15.6 N	3	US astronomer (1872–1942)
38	Daly	Apollonius A	56.8 W	5.0 N	24	Us geologist (1871–1957)
23	Daubrée	Menelaus S	14.8 W	15.7 N	14/1590	French geochemist (1814–96)
26	Eckert		58.4 W	17.8 N	3	US astronomer (1902–71)
25	Franck	Römer K	35.6 W	22.7 N	12/2510	German–US physicist (1882–1964)
18	Freud		52.3 E	25.8 N	2	Austrian psychoanalyst (1856–1939)
22	Galen	Aratus A	5.1 W	22.0 N	10	Greek physician (2nd century AD)
22	Hadley	Hadley C	2.9 W	25.4 N	6/1120	UK astronomer (1682–1744)
49	Haldane		84.0 W	1.7 S	35	UK physiologist (1860–1936)
25	Hill	Macrobius B	40.8 W	21.2 N	16/3340	US astronomer (1838–1914)
23	Hornsby	Aratus CB	12.4 W	23.8 N	3	UK astronomer (1733–1810)
49	Houtermans		87.0 W	9.3 S	35	German–Swiss physicist (1902–66)
8	Humason	Lichtenberg G	56.7 E	30.7 N	4	US astronomer (1891–1972)
22	Huxley	Wallace B	4.6 E	20.2 N	4/840	UK zoologist (1825–95)
23	Joy	Hadley A	6.6 W	25.1 N	6/1000	US astronomer (1882–1973)
49	Kiess		84.0 W	6.3 S	65	US spectroscopist (1887–1967)
38	Knox-Shaw	Banachiewicz F	80.3 W	5.3 N	12	UK astronomer (1885–1970)
49	Kreiken		84.5 W	9.0 S	15	Dutch astronomer (1896–1964)

Map number	Crater	Old name	Selenographic longitude	latitude	Diameter (km) Depth (m)	Personality
37	Lawrence	Taruntius M	43.3 W	7.6 N	24/1000	US physicist (1901–58)
25	Lucian	Maraldi B	36.8 W	14.6 N	7/1490	Greek author (2nd century AD)
8	Nielsen	Wollaston C	51.9 E	31.8 N	10	Danish astronomer (1902–70)
38	Peek		87.1 W	2.8 N	12	UK amateur astronomer (1891–1965)
49	Runge		86.8 W	2.3 S	35	German mathematician/ physicist (1856–1927)
24	Sarabhai	Bessel A	21.0 W	24.7 N	8/1660	Indian physicist (20th century)
38	Shapley	Picard H	56.8 W	9.9 N	23	US astronomer (1885–1972)
22	Spurr	Archimedes M	3.2 E	26.1 N	3.5/570	US geologist, selenologist (1870–1950)
38	Tacchini	Neper K	86.1 W	5.2 N	39	Italian astronomer (1838–1905)
37	Tebbutt	Picard G	53.7 W	9.5 N	32	Australian amateur astronomer (1834–1916)
25	Theo-phrastos	Maraldi M	39.1 W	17.5 N	9/1700	Greek philosopher (4th century BC)
18	Väisälä	Aristarchus A	47.8 E	25.9 N	8	Finnish astronomer, geodesist (1891–1971)
24	Very	Le Monnier B	25.3 W	25.7 N	5/950	US astrophysicist (1852–1927)
37	Watts	Taruntius D	46.3 W	9.0 N	15	US astricist (1889–1971)
49	Widmannstätten		85.5 W	6.0 S	50	Austrian meteoricist (c. 1753–1849)
22	Yangel	Manilius F	4.7 W	17°.0 N	8	Soviet space scientist (d. 1971)
18	Zinner	Schiaparelli B	58.7 E	26.7 N	3.5	German historian of astronomy (1886–1970)

INDEX TO MAPS OF THE MOON

(Alphabetical list of names with numbers of the maps on which the relevant formation is found.)

The Roman numerals refer to the general six-part map, the Arabic numerals refer to the detailed map in 76 plates.
The numbers in brackets refer to maps which show only a part of the given formation.

CRATERS

Abbe III
Abbot 37
Abel 69
Abenezra 56
Abulfeda 45, (56)
Adams 69
Agatharchides 52
Agrippa 34
Airy (55), 56
Aitken III
Albategnius I, 44, (45)
Aldrin 35
Alexander 13
Alfraganus 46
Alhazen 27
Aliacensis 55, 65
Almanon 56
Alpetragius 55
Alphonsus I, 44, (55)
Amundsen VI, 73
Anaxagoras V, 4
Anaximander 2
Anaximenes 3
Anděl I, (45)
Anderson III
Angström 9, 19
Ansgarius II, 49
Antoniadi VI
Apianus 56
Apollo III
Apollonius 38
Appleton III
Arago I, 35
Aratus 22
Archimedes I, 12, 22
Archytas I, V, 4
Argelander (55), 56
Ariadaeus 35
Aristarchus IV, 18
Aristillus I, 12
Aristoteles I, V, 5
Armstrong 35
Arnold V, 5
Arrhenius IV
Arzachel I, 55
Asclepi 74

Aston 8
Atlas I, II, (7), 15
Autolycus 12, (22)
Auwers 24, (35)
Auzout 38
Avogadro V
Azophi I, 56

Baade IV, 61
Babbage V, 2
Baco I, VI, 74
Baillaud V, 5
Bailly VI, 71
Baily 6
Balboa 17
Ball 64
Balmer 60
Banachiewicz 38
Banting 23
Barnard 60, (69)
Barocius I, 66
Barrow 4
Bartels 17
Bayer 71
Beaumont (57), 58
Bečvář II
Beer 21
Behaim 60
Belkovich V, 7
Bellot 48
Bernouilli 16
Berosus 16
Berzelius 15
Bessarion 19
Bessel I, (23), 24
Bettinus VI, 71
Bhabha III, VI
Bianchini 2, 10
Biela I, II, VI, 75, (76)
Billy 40, (51)
Biot 59
Birkhoff III, V
Birmingham 4
Birt I, 54
Blagg 33
Blancanus VI, 72

Blanchinus 55
Bode 33
Boguslawsky 74
Bohnenberger 58
Bohr 28
Bond, G. 15
Bond, W. V, 4
Bonpland 42
Boole V, 2
Borda 59
Boscovich 34
Bose III, VI
Boss 16
Bouguer 2
Boussingault VI, 74, 75
Bowen 23
Brackett 24
Brayley I, 19
Breislak 66
Brenner 68
Brianchon V, 2, 3
Briggs 17
Brisbane 76
Brouwer IV
Brown 64
Bruce 33
Buch 66
Bulliadus I, 53
Bunsen 8
Burckhardt 16
Bürg 14
Burnham 45
Büsching 66, (67)
Byrd V, 4
Byrgius 50, (51)

Cabannes VI
Cabeus 73
Cajal 36
Calippus 13
Cameron 37
Campanus 53
Campbell III
Cannizzaro IV
Cannon 27
Capella I, 47

245

Capuanus I, 63
Cardanus 28
Carlini 10
Carmichael 25
Carnot III, V
Carpenter V, 2, 3
Carrington 15
Casatus VI, 72
Cassini I, 12
Catalán 61
Catharina I, 57
Cauchy I, II, 36
Cavalerius 28
Cavendish 51
Cayley 34
Celsius 67
Censorinus 47
Cepheus 1
Chacornac 14, (24), (25)
Challis 4
Chaplygin III
Chappell V
Chebyshev III, IV
Chevallier 15
Chladni 33
Cichus 63
Clairaut 66
Clausius 62
Clavius I, VI, 72, (73)
Cleomedes II, 26
Cleostratus 1
Clerke 25
Collins 35
Colombo (48), 59
Compton II, V
Condorcet 38
Conon I, 22
Cook 59
Copernicus I, 31
Coulomb III, IV, V
Cremona 2
Crozier 48
Crüger 50
Curie II
Curtis 26
Curtius VI, 73
Cusanus V, 6
Cuvier I, 74
Cyrillus I, 46, (57)
Cysatus 73

Daedalus III
Daguerre 47
D'Alembert III, V
Dalton 17
Daly 38
Damoiseau 39
Daniell 14
Darney 42, 53

d'Arrest 34
Darwin IV, 50
Daubrée 23
da Vinci 37
Davy 43
Dawes 24
Debes 26
Debye III
Dechen 8
de Gasparis 51
Delambre 46
de la Rue I, II, V, 6
Delaunay 55
Delisle 9, 19
Delmotte 26
Deluc 73
Dembowski 34
Democritus 5, 6
Demonax VI, 74
de Morgan 34
Desargues 2
Descartes 45
Deseilligny 24
Deslandres I, (64), 65
de Vico 50, 51
Dionysius 35
Diophantus 19
Dollond 45
Donati 55
Doppelmayer 52
Dove 67
Draper 20
Drebbel 62
Drygalski VI, 72
Dubiago 38
Dunthorne 53, 62, 63
Dyson V

Eckert 26
Eddington IV, 17
Egede 5, 13
Eichstädt 50
Eimmart 27
Einstein IV, 17
Elger 63
Encke 30
Endymion II, V, 7
Epigenes 4
Epimenides 63
Eratosthenes 21, 32
Euclides I, 41
Euctemon 5
Eudoxus I, 13
Euler (19), 20

Fabricius 68
Fabry II
Faraday 66

Fauth 31
Faye 55
Fechner VI
Fermat 57
Fermi II
Fernelius 65
Fersman IV
Feuillée 21
Firmicus 38
Fitzgerald III
Fizeau III, IV, VI
Flammarion 44
Flamsteed I, 40
Fleming II
Fontana 51
Fontenelle 3
Foucault 2
Fourier 51, 61
Fowler III
Fracastorius I, 58
Fra Mauro I, 42, (43)
Franck 25
Franklin 15
Franz 25
Fraunhofer 69
Freud 18
Fridman III, IV
Furnerius II, 69

Gagarin III
Galen 22
Galilaei IV, 28
Galle 5
Galois III
Galvani IV, 1, 8
Gambart (31), 32
Gamow V
Gärtner V, 6
Gassendi I, 52
Gaudibert 47
Gauricus 64
Gauss II, 16
Gay-Lussac 31
Geber 56
Geminus 16
Gemma Frisius 66
Gerard 8
Gibbs 60
Gilbert 49
Gill 75
Gioja 4
Glaisher 37
Goclenius 48
Goddard 27
Godin 34
Goldschmidt 4
Goodacre 66
Gould 53, (54)
Graff 61

Grimaldi   IV, 39
Grove   14
Gruemberger   73
Gruithuisen   9
Guericke   43
Gum   69
Gutenberg   48
Gyldén   44

Hadley   22
Hagecius   VI, 75
Hahn   16, (27)
Haidinger   63
Hainzel   63
Haldane   49
Hale   VI, 74, 75
Hall   15
Halley   44, 45
Hamilton   69
Hanno   76
Hansen   27, 38
Hansteen   40
Harding   8
Harpalus   I, V, (1), 2
Hartwig   39
Hase   59
Hausen   VI, 71
Hayn   V, 6
Heaviside   III
Hecataeus   60
Hedin   28
Heinsius   64
Heis   10
Helicon   10
Hell   64
Helmholtz   VI, 75
Henry, Paul   51
Henry, Prosper   51
Heraclitus   73
Hercules   I, II, (6), 14
Herigonius   41
Hermann   39
Hermite   V, 4
Herodotus   IV, 18
Herschel, C.   10
Herschel, J.   V, 2
Herschel, W.   44
Hertzsprung   III, IV
Hesiodus   54
Hess   III, VI
Hevelius   28
Hilbert   II
Hill   25
Hind   45
Hippalus   52, (53)
Hipparchus   I, 44, 45
Holden   60
Hommel   I, VI, 75
Hooke   15

Hornsby   23
Horrebow   2
Horrocks   (44), 45
Hortensius   30, (31)
Houtermans   49
Hubble   27
Huggins   65
Humason   8
Humboldt   II, 60
Huxley   22
Hyginus   I, 34
Hypatia   46

Icarus   III
Ideler   74
Inghirami   IV, (61), 62
Isidorus   47

Jacobi   I, 74
Jansen   (35), 36
Jansky   38
Janssen   I, II, 67, 68
Jeans   VI
Joffe   III, IV
Joliot   II
Joy   23
Jules Verne   III
Julius Caesar   34

Kaiser   65, (66)
Kane   5
Kant   46
Kapteyn   49
Karpinsky   V
Kästner   49
Keeler   III
Kepler   I, 30
Kies   53
Kiess   49
Kinau   74
Kirch   12
Kircher   71, (72)
Kirchhoff   15
Klaproth   VI, 72
Klein   44
Koch   III
König   53
Korolev   III
Krafft   17
Krasnov   50, (61)
Kreiken   49
Krieger   19
Krusenstern   55, (56)
Kugler   VI
Kulik   III
Kunowsky   30
Kurchatov   II, III

La Caille   55
La Condamine   2
Lacroix   61
Lade   45
Lagalla   63
Lagrange   IV, 61
Lalande   I, 43
Lamarck   50
Lambert   20
Lamé   II, 49, 60
Lamèch   13
Lamont   35
Landau   IV
Langley   1
Langmuir   III, IV
Langrenus   II, 49
Lansberg   (30), (31), (41), 42
La Pérouse   49
Lassel   (43), 54
Lavoisier   8
Lawrence   37
Lebedev   II
Lee   62
Legendre   (59), 60
Le Gentil   VI, 72
Lehmann   61, 62
Leibnitz   III
Lemaitre   VI
Le Monnier   24, 25
Lepaute   62
Letronne   40
Le Verrier   I, 10, 11
Lexell   65
Liapunov   27
Licetus   65
Lichtenberg   8
Lick   37
Liebig   51
Lilius   73
Lindenau   67
Linné   I, 23
Lippershey   54
Littrow   25
Lockyer   67
Loewy   52
Lohrmann   39
Lohse   49
Lomonosov   II
Longomontanus   I, (64), 72
Lorentz   IV
Louville   9
Lubbock   48
Lubiniezky   53
Lucian   25
Luther   14
Lyell   36
Lyman   VI

247

Lyot II, VI, (69), 76

Mach III
Maclaurin 49
Maclear 35
Macrobius 26
Mädler 47
Maestlin 29
Magelhaens 48
Maginus I, VI, (65), (72), 73
Main 4
Mairan I, 9
Malapert 73
Mallet 68
Manilius 23, 34
Manners 35
Manzinus VI, 74
Maraldi 25
Marco Polo 22, (33)
Marinus 69
Marius IV, 29
Markov IV, V, 1
Marth 63
Maskelyne 36
Mason 14
Maupertuis 2, (3)
Maurolycus I, 66
Maury 15
Mayer, C. 5
Mayer, T. 19
Mc Clure 59
Mee I, 63
Mendel IV
Mendeleev II, III
Menelaus 23, (24)
Mercator 53
Mercurius II, 15, (16)
Mersenius IV, 51
Messala II, 16
Messier I, II, 48
Metius 68
Meton V, 4, 5
Michelson IV
Milichius 30
Miller 65
Millikan II
Milne II
Minkowski III, VI
Minnaert VI
Mitchell 5
Moigno 5
Moltke (35), 46
Monge 59
Montanari 64
Moretus VI, 73
Morse III
Mösting (32), 43, (44)
Mouchez 3, 4

Müller 44
Murchison 33
Mutus VI, 74

Nansen V, 5
Nasireddin 65
Nasmyth 70
Naumann 8
Neander 68
Nearch 75
Neison 5
Neper II, 38
Neumayer 75
Newcomb (15), 25
Newton VI, 73
Nicholson 50
Nicolai 67
Nicollet 54
Nielsen 8
Nöggerath 70
Nonius 65
Nöther V
Numerov VI
Nušl III

Oenopides 1
Oersted 15
Ohm IV
Oken II, 69
Olbers 28
Opelt 53
Oppenheimer III
Oppolzer 44
Orontius 65

Palisa 43
Palitzsch 59
Pallas 33
Palmieri 51
Parrot (44), 55
Parry 42, 43
Pascal V, 3
Pasteur II
Pauli II, III
Pavlov II, III
Peary V, 4
Peek 38
Peirce 26
Peirescius 68, (69)
Pentland 73
Petavius II, 59
Petermann 5
Peters 5
Petrov 76
Pettit 50
Petzval VI
Phillips 60
Philolaus V, 3
Phocylides IV, VI, 70

Piazzi IV, 61
Piazzi Smyth 12
Picard 26, 37
Piccolomini (57), 58, (68)
Pickering 45
Pictet 64, (65)
Pingré IV, VI, 70
Pitatus 54, (64)
Pitiscus VI, 75
Plana 14
Planck II, III, VI
Plaskett V
Plato I, V, 3, 4
Playfair 56
Plinius 24, (35)
Plutarch II, 27
Poincaré III, VI
Poisson (56), 66
Polybius 57
Poncelet V, 3
Pons 57
Pontanus 56
Pontécoulant II, VI, 76
Porter 72, (73)
Posidonius I, 14
Poynting III, IV
Priestly II
Prinz 19
Proclus 26
Proctor 65
Protagoras 4
Ptolemaeus I, 44
Puiseux 52
Purbach 55
Pythagoras V, 2
Pytheas I, 20

Rabbi Levi I, 67
Ramsden 63
Rayleigh 27
Réaumur 44
Regiomontanus 55
Régnault 1
Reichenbach (59), (68), 69
Reimarus 68
Reiner 28, 29
Reinhold V, 1
Repsold V, 1
Rhaeticus 33, (34), (44), (45)
Rheita I, II, 68
Riccioli IV, 39
Riccius 67
Riemann 16
Ritchey 45
Ritter 35
Roberts V
Robertson IV

Robinson 2
Rocca 39
Roche II, III
Römer I, 25
Röntgen IV
Rosenberger 75
Ross 35
Rosse 58
Rost 71
Rothmann 67
Rowland III, V
Rozhdestvensky V
Runge 49
Russel 17
Rutherfurd 72

Sabine 35
Sacrobosco I, 56, (57)
Santbech 59
Sarabhai 24
Sasserides 64
Saunder 45
Saussure 65
Scheiner VI, 72
Schiaparelli 18
Schickard IV, 62
Schiller I, VI, 71
Schlütter 39
Schmidt 35
Schomberger (73), 74
Schorr 60
Schrödinger VI
Schröter 32
Schubert 38
Schumacher 16
Schwabe 6
Schwarzschild V
Scoresby 4
Scott VI, (73), 74
Seares V
Secchi 37
Seeliger 44
Segner 71
Seleucus 17
Seneca 27
Seyfert II
Shaler 61
Shapley 38
Sharp 9, 10
Sheepshanks 5
Short 73
Shuckburgh 15
Silberschlag 34
Simpelius 73
Sinas 36
Sirsalis 39
Sklodowska II
Smoluchowski V

Snellius 59, (69)
Sommerfeld V
Sömmering 32, 43
Sosigenes 35
South IV, V, 2
Spallanzani 67
Spörer 44
Spurr 22
Stadius I, 32
Stebbins V
Stefan IV
Steinheil (68), 76
Steklov IV
Stevinus 69
Stiborius 67
Stokes 1
Stöfler I, 65, (66)
Störmer III, V
Strabo V, 6
Street 64
Struve IV, 17
Suess 29
Sulpicius Gallus 23
Sylvester 3

Tachinni 38
Tacitus 57
Tacquet 24
Tannerus 74
Taruntius I, II, 37
Taylor (45), 46
Tebbutt 37
Tempel 34
Thales 6
Theaetetus 12
Thebit 55
Theon Junior 45, 46
Theon Senior 45, (46)
Theophilus I, 46, (47)
Theophrastos 25
Tikhov V
Timaeus 4
Timocharis I, 21
Tisserand 26
Torricelli (46), 47
Tralles 26
Triesnecker I, 33
Trouvelot 4, 12
Trumpler III
Tsiolkovskij II, III
Turner 43
Tycho I, 64

Ukert 33
Ulugh Beigh 8

Väisälä 18
Van de Graaff III

Vasco da Gama 28
Vavilov III, IV
Vega 68
Vendelinus II, (49), 60
Vernandsky II, III
Vieta IV, 51, (61)
Vitello (52), 62
Vitruvius 25
Vlacq 75
Vogel 45, 56
Volta IV, V, 1
Von Kármàn III
Voskresensky 17

Wallace 21
Walter I, 65
Wargentin IV, 70
Watt I, II, 76
Watts 37
Webb (38), 49
Weigel 71
Weinek 58
Weiss 63
Wells, H. G. II
Werner 55
Wexler 75
Whewell 34
Wichmann 41
Widmännstätten 49
Wiener III
Wilhelm I, (63), 64
Wilkins 57
Williams 14, (15)
Wilson 72
Wöhler 67
Wolf 54
Wollaston 9
Wright 61
Wrottesley 59
Wurzelbauer 64

Xenophanes IV, V, 1

Yablochkov V
Yangel 22
Yerkes 26, 37
Young 68

Zach VI, 73
Zagut 67
Zeeman VI
Zeno 16
Zinner 18
Zöllner 46
Zsigmondy V
Zucchius 71
Zupus 51

MOUNTAINS (MONS)

Ampère 22	Hadley 22	Piton 12
Argaeus 24	Huygens 22	Rümker IV, 8
Blanc 12	La Hire 20	Serao 21
Bradley 22	Pico 11	Wolff 21

MOUNTAIN RANGES (MONTES) AND FAULTS (RUPES)

Montes Alpes I, 4, 12
Montes Apenninus I, 21, 22
Montes Carpatus I, 20, 31
Montes Caucasus 13
Montes Cordillera IV, 39, 50
Montes Haemus 23
Montes Harbinger 19
Montes Jura 2
Montes Pyrenaeus 48, 58
Montes Recti 11

Montes Riphaeus 41, 42
Montes Rook IV, 50
Montes Spitzbergensis (11), 12
Montes Taurus I, II, 25
Montes Teneriffe 11

Rupes Altai I, 57
Rupes Cauchy 36
Rupes Liebig 51
Rupes Recta 54

CAPES (PROMONTORIUM)

Prom. Agarum 38
Prom. Agassiz 12
Prom. Archerusia 24
Prom. Banat 20
Prom. Deville 12
Prom. Fresnel 22

Prom. Heraclides 10
Prom. Kelvin 52
Prom. Laplace 10
Prom. Lavinium 26
Prom. Olivium 26
Prom. Taenarium 54

VALLEYS (VALLIS)

Vallis Alpes 4, 12
Vallis Baade 61
Vallis Bouvard 61, IV
Vallis Inghirami 61

Vallis Palitzsch 59
Vallis Rheita 68
Vallis Schröteri 18
Vallis Snellius 59, 69

RILLES AND CLEFTS (RIMA)

Rima Ariadaeus 34
Rimae Aristarchus 18
Rima Bradley 22
Rimae Bürg 14
Rima Cardanus 28
Rimae Daniell 14
Rimae Fresnel 22
Rimae Goclenius 48
Rima Hadley 22

Rima Hesiodus 53, 63
Rimae Hippalus 52, 53
Rima Hyginus 33, 34
Rimae Hypatia 35, 46
Rima Krafft 17
Rimae Littrow 25
Rima Marius 18
Rima Sirsalis 39, 50
Rimae Triesnecker 33

LAKES, MARSHES AND BAYS

Lacus Aestatis 39, 50
Lacus Autumnae 39, 50
Lacus Mortis 14
Lacus Somniorum 14
Lacus Veris 39, 50
Palus Epidemiarum 53, 63

Palus Putredinis 22
Palus Somni 26, 37
Sinus Aestuum I, 32, 33
Sinus Iridum I, 10
Sinus Medii I, 33, 44
Sinus Roris IV, V, 1, 9

SEAS (MARE)

Mare Anguis   27
Mare Australe   II, VI, 69, 76
Mare Cognitum   I, 42
Mare Crisium   II, 26, 27, 37, 38
Mare Fecunditatis   I, II, 37, 38, 48, 49, 59
Mare Frigoris   I, V, 2, 3, 4, 5, 6
Mare Humboldtianum   II, V, 7
Mare Humorum   I, 51, 52
Mare Imbrium   I, 9, 10, 11, 12, 19, 20, 21
Mare Ingenii   III
Mare Marginis   II, 27, 38
Mare Moscoviense   III
Mare Nectaris   I, 47, 58
Mare Nubium   I, 53, 54
Mare Orientale   IV, 50
Mare Serenitatis   I, 13, 14, 23, 24
Mare Smythii   II, 38, 49
Mare Spumans   38
Mare Tranquillitatis   I, 35, 36, 37, 47
Mare Undarum   38
Mare Vaporum   I, 22, 33, 34
Oceanus Procellarum   I, IV, 8, 9, 17, 18, 19, 28, 29, 30, 39, 40, 41

COMMEMORATIVE NAMES

Planitia Descensus   28
Sinus Lunicus   12
Statio Tranquillitatis   35

INDEX TO MAPS OF MARS

(Alphabetical list of names with the numbers of the maps on which the relevant formation is found.)

ALBEDO FORMATIONS (printed black on maps)

Abalos E
Achillis Fons B2a
Achillis Pons A1c,d
Aeolis C3d
Aeria A2a; D2d
Aetheria D1a,b
Aethiopis D2a
Alba B1c
Amazonis C2b
Amenthes D2b
Anian C1d; D1a
Antigones Fons D2c
Aonius Sinus B4c
Arabia A2a,b; D2d
Aram A2c
Araxes B3d
Arcadia B1d; C1a
Arethusa Lacus A1b; E
Argenteus Mons F
Argus A2c
Argyre I A4d; B4a; F
Argyre II F
Arsia Silva B3d
Ascraeus Lacus B2c
Astusapes D1c,d
Atlantis C4b
Aurorae Sinus A3d; B3a
Ausonia D4a, b

Baetis B3a
Baltia E
Bathys B4c
Boreosyrtis D1c,d
Bosporos B4b

Callirrhoes Sinus A1c
Candor B2b
Casius D1b,c
Castorius Lacus C1b; E
Cebrenia C1c,d
Cecropia E
Cerberus C2c
Chalce A4c
Chaos C1c
Chersonesus D4b; F
Chryse A2c,d
Chrysokeras B4c
Claritas B3c
Coloe Palus D1d

Copais Palus E
Coprates B3b
Crocea D2c
Cyane B2d; C2a
Cyclopia C2d; D2a
Cydonia A1b

Daedalia B3c,d
Deltonon Sinus D3d
Depressio Magna F
Depressio Parva F
Deucalionis Regio A3b
Deuteronilus A1b,c
Dia F
Diacria C1b
Dioscuria A1a; D1d; E
Eden A2b
Edom A3b
Electris C4b
Elysium C2c,d
Eos A3d
Erebus C2c
Eridania C4c, d
Eunostos C2d; D2a
Euripus D4c; F
Euxinus Lacus C1b

Fastigium Aryn A3b

Ganges B2a
Gehon A2b
Gigantum Sinus C3b
Gomer Sinus D2,3a
Gorgonum Sinus C3b

Hammonis Cornu A3a; D3d
Hecates Lacus C1c
Hellas D4c,d; F
Hellespontica Depressio F
Hellespontus A4a
Herculis Pons C1c
Hesperia D3a,b
Hyblaeus C2d; D2a
Hydrae Palus A2d; B2a
Hydrates A2c
Hyperboreus Lacus E

Iapygia D3c
Icaria B4d

252

Idaeus Fons   A1d; B1a
Ierne   E
Indus   A2c
Isidis Regio   D2c
Ismenius Lacus   A1b
Jamuna   A2d; B2a
Juventae Fons   B3b

Laestrygonum Sinus   C3c
Lemuria   E
Libya   D2b
Lunae Lacus   B2b
Lycus   B1d; C1a

Mare Acidalium   A1c,d; E
Mare Amphitrites   F
Mare Australe   F
Mare Boreum   E
Mare Chronium   F
Mare Cimmerium   C3d; D3a
Mare Erythraeum   A3c,d; B3a
Mare Hadriacum   D3c; D4b
Mare Ionium   A3a; D3d
Mare Oceanidum   A4d; B4a; F
Mare Serpentis   A3a
Mare Sirenum   C3b; C4a,b
Mare Tyrrhenum   D3b
Margaritifer Sinus   A3c
Melas Lacus   B3b
Memnonia   C3a
Meotis Palus   E
Meridiani Sinus   A3b,c
Meroe   D1c
Mesogaea   C2b
Moab   A2b
Moeris Lacus   D2c

Nectar   B3b
Neith Regio   D1b,c
Nepenthes-Thoth   D2b
Nereidum Fretum   B4a
Niliacus Lacus   A2c,d
Nilokeras   A1d; B1a
Nilosyrtis   D2c
Nix Cydonia   A1c
Nix Olympica   B2d; C2a
Noachis   A4b,c
Nodus Alcyonius   D1b
Nodus Gordii   B3d; C3a
Nodus Laocoontis   D2b
Novus Mons   F
Nubis Lacus   D2b

Ogygis Regio   B4b; F
Olympia   E
Ophir   B3b
Ortygia   E
Oxia   A2c
Oxia Palus   A2c
Oxus   A2c

Palinuri Fretum   C4a,b; F
Panchaia   E
Pandorae Fretum   A3b
Pavonis Lacus   B2c
Phaethontis   C4a
Phlegra   C2c
Phoenicis Lacus   B3c
Pierius   E
Pontica Depressio   B4b
Promethei Sinus   F
Propontis I   C1b
Propontis II   C1b,c; E
Protei Regio   A3d; B3a
Protonilus   A1a
Pyrrhae Regio   A3c

Rasena   C3c

Sabaeus Sinus   A3a,b
Scamander   C4c
Scandia   E
Sigeus Portus   A3b
Sinai   B3b
Sirenum Sinus   B4d; C4a
Sithonius Lacus   C1d; D1a; E
Solis Lacus   B3b
Styx   C2c
Syria   B3b,c
Syrtis Major   D2c
Syrtis Minor   D3b

Tanais   A1d; B1a; E
Tempe   B1b; E
Tharsis   B2b,c, d
Thaumasia   B4b
Thoana Palus   D1b
Thyle I   F
Thyle II   F
Thyles Collis   F
Thyles Mons   F
Thymiamata   A2c
Tiphys Fretum   C4d; D4a; F
Titanum Sinus   C3b
Tithonius Lacus   B3b,c
Trinacria   D3b, c
Trivium Charontis   C2c

Uchronia   E
Umbra   D1c; E
Utopia   D1b; E

Vulcani Pelagus   A4c,d

Xanthe   A2d; B2a
Xanthes   C4d; D4a

Yaonis Regio   A4a; D4d

Zephyria   C2,3c

253

CRATERS ON MARS (printed red on maps)

Adams C1c
Antoniadi D2c
Arago A2a

Baldet D2c
Barnard F
Becquerel A2c
Boeddicker C3c
Bond A4d; B4a
Brashear B4c
Burton C3b

Campbell C4c; F
Cassini A2a
Cerulli A1b
Clark B4d; C4a; F
Coblentz B4b; F
Columbus C3b
Copernicus C4b
Cruls C4c
Curie A2c

Darwin A4c; F
Dawes A3a

Eddie C2d; D2a

Fessenkov B2b
Flammarion A2a; D2d
Flaugergues A3b
Focas A1b
Fourier D3c

Gale C3d; D3a
Galle A4d; F
Gill A2b
Gledhill D4c
Graff C3c
Green A4c; F

Hale A4d; B4a
Halley B4a
Hartwig A4c
Helmholtz A4c
Henry A2b
Herschel C3d; D3a
Holden A3d; B3a
Holmes F
Hooke A4d; B4a
Huggins C4c
Hussey B4d; C4a; F
Huygens D3d

Janssen A2a
Jarry-Desloges D3c

Kaiser A4b
Kepler C4d; D4a
Knobel C3d; D3a
Korolev E
Kunowsky A1c; E

Lamont B4c
Lassel B3b
Lau F
Le Verrier A4b
Li Fan C4b
Liu Shin C4b; F
Lockyer C2c
Lohse A4c
Lomonosov E
Lowell B4b; F
Lyot A1b; E

Maggini A2b
Martz C4d; D4a
McLaughlin A2c
Mendel C4c; F
Mie C1d; D1a
Milankovitch C1a; E
Millochau D3c
Molesworth C3d
Moreux A1a; D1d
Müller C3d; D3a

Newcomb A3b
Newton C4b
Nicholson C3b
Niesten D3d

Oudemans B3c

Pasteur A2b
Perepelkin B2c
Péridier D2c
Pettit C2b
Phillips F
Pickering B4d; C4a
Porter B4c
Priestley C4d; D4a
Proctor A4b

Quénisset A1a; D1d

Rabe A4a
Rayleigh F
Renaudot D1c
Ritchey A3d; B3a
Ross B4c; F
Rudaux D1d
Russel A4b; F

254

Schaeberle   D3d
Schiaparelli   A3b
Schmidt   F
Schröter   D3d
Secchi   D4b; F
Slipher   B4b
South   F
Spallanzani   D4c
Stokes   C1c; E
Stoney   F

Teisserenc   A3a
Terby   D3c
Tikhov   D4b; F
Tycho Brahe   C4d; D4a
Tyndall   C1c

Vogel   A4c

Wallace   D4b
Wells   F
Williams   C3b
Wright   C4b

Acknowledgments

This book would not have been completed were it not for the sympathetic consideration and valuable cooperation of a number of specialists, who have allowed the author access to essential information and data in order to compile the text and chart sections. Foremost among these have been E. A. Whittaker of the Lunar and Planetary Laboratory in Tucson, USA; Professor Z. Kopal of Manchester University, England; Dr. T. W. Rackham of the Jodrell Bank Observatory in England, C. A. Wood and D. W. G. Arthur, prominent American selenographers and Dr. L. D. Jaffe of the Jet Propulsion Laboratory, Pasadena, USA. The author extends his sincere thanks to all of them. Finally, last but not least, the author extends his thanks to his wife for the invaluable help and moral support she gave him during the two years of work on this atlas.